I dedicate this book to Jeanine,
my wife and true soulmate,
and to our three beautiful and healthy girls,
Rachel, Kelsey and Megan.
I love you, forever and always,
Dad

THE DIABETES CURE

The Diabetes Cure

*How you can slow, stop and
even cure type 2 diabetes*

**Dr Vern S. Cherewatenko
and Paul Perry**

Thorsons
Directions for Life

While the authors of this work have made every effort to ensure that the information contained in this book is as accurate and up to date as possible at the time of publication, medical and pharmaceutical knowledge is constantly changing and the application of it to particular circumstances depends on many factors. Therefore it is recommended that readers always consult a qualified medical specialist for individual advice. This book should not be used as an alternative to seeking specialist medical advice, which should be sought before any action is taken. The authors and publishers cannot be held responsible for any errors and omissions that may be found in the text, or any actions that may be taken by a reader as a result of any reliance on the information contained in the text, which is taken entirely at the reader's own risk.

Thorsons
An Imprint of HarperCollins*Publishers*
77–85 Fulham Palace Road,
Hammersmith, London W6 8JB

The Thorsons website address is: www.thorsons.com

First published by Cliff Street Books,
an imprint of HarperCollins*Publishers*, 1999

This Thorsons edition published 2000

1 3 5 7 9 10 8 6 4 2

A catalogue record for this book
is available from the British Library

ISBN 0 7225 3924 X

Printed and bound in Great Britain by
Creative Print and Design (Wales), Ebbw Vale

Contents

Acknowledgements

I would like to acknowledge the following people for assisting me to make this book more than just a vision.

My supportive wife, Jeanine, who bears the added responsibility of having a physician for a spouse, you are the wind beneath my wings. My children, Rachel, Kelsey and Megan for all your hugs and finding the time to help me in so many ways. My co-author, Paul Perry, a pillar of literary excellence. To Mel Morse, MD, for introducing me to Paul and the staff of HarperCollins Publishers. To Diane Reverand, Editor and Publisher of Cliff Street Books, who endorsed the vision of helping educate people about such a severe disease. To Krista Stroever, who keeps it all happening, on time and on schedule, and Andrea Molitor, production editor, for her endless patience. To our agent, Nat Sobel, and his staff for handling the details. We also want to thank Mardy Fones for her editorial guidance. Without her fine work, this book would have taken far longer to produce.

A special thanks to my partner and colleague, David MacDonald, DO, who has provided unyielding inspiration and assistance. To Dr Sohelya Radfar, who has given so much of herself to teach a better way of living to me and others. To my dear friend James Coffin, for illustrating my life and providing wisdom's pearls. My medical colleagues, Kevin Connolly, PhD, C J DelMarco, DO, Pieter de Wet, MD, Steve Kaye, MD, Chris Renna, DO, Fred Miser, MD, Joan Robnett, Steve Hughes, MD,

Bev Kent, RD, CD, Nass Ordoubadi, MD, Andy Abolins, MD, Anthony Conte, MD, Bill Downs at Interhealth, Howard Lyman, Jeff Bland, PhD, at HealthComm, for providing your medical expertise.

The strength and patience behind the scenes: Michael Pennington – 'Superman' and business associates Larry Ream, Gregg Jordshaugen, Elaine Wilcox and Ron Bozi. A special thank you to Gene Johnson, who went above and beyond. To Gwen Hoeck, my first employee, who has remained with me since residency. To the dedicated office staff, who has remained at my side, through all the challenges that face health care practitioners today. To my patients, who have remained flexible with my ever-changing schedule, your patience is a virtue. To those true friends who ... no matter how long between talks ... make me feel like I'm 'home' again when we get together.

Last, but not least, to the patients who have diabetes mellitus, and to the dedicated researchers, organizations, doctors and educators who are working towards curing this destructive disease.

Thank you all and please enjoy ... *The Diabetes Cure*!

What *The Diabetes Cure* is About ...

The Diabetes Cure provides the knowledge for understanding diabetes which can lead to the control – and even the cure – of one of the most devastating illnesses of our time.

The Diabetes Cure will provide you with the necessary information and tools to halt and even reverse the effects of type 2 (adult-onset) diabetes. It will also introduce you to a herb with the chemical name 'hydroxycitric acid', and the mineral chromium, both of which can improve your body's response to insulin without your having to take some of the 'harder' glucose-control drugs.

The Diabetes Cure provides important warning signs and medical history indications to prevent diabetes in those determined to be at highest risk.

The Diabetes Cure is a fresh and aggressive look at the ultimate cause of the disease of diabetes and a natural approach to curing this disease, one that has been traditionally *managed* by the medical profession.

Through *The Diabetes Cure*, you will learn and understand the mechanisms which cause type 2 diabetes and a natural approach to become *un-diabetic* from a programme implemented by a qualified GP.

Disclaimer

Diabetes is a complicated and devastating disease. The material presented in this book is for patient education and informational purposes only. It is not meant to be a substitute for professional medical advice.

Diabetes is a disease which should be treated by a health care professional or a team of health care professionals with advanced training in diabetes.

Please discuss all aspects of *The Diabetes Cure* with your health care provider before using any of the information given here. It is important that you seek consultation with your health care provider before you change or alter any of the medications you may be taking.

The Diabetes Cure is a response to the request for a non-prescription, or *natural*, approach to diabetes mellitus, a disease that affects nearly 2 million people in the United Kingdom.

The evidence supports the conviction that increased understanding of diabetes combined with an aggressive nutrition and exercise programme enhanced by nutritional supplements can have a pronounced and positive effect on those with type 2 diabetes mellitus.

The authors trust that the information in *The Diabetes Cure* will give patients a greater understanding of diabetes and the ability to increase compliance with a supervised medical programme to control and cure type 2 diabetes. To *'CURE'* is to apply a treatment, a drug or another method to treat a medical problem. It means to restore the health of a patient with a disease or disorder or obtain a favourable result by treating a disease or disorder. There is no overt or implied guarantee that the information presented here will benefit those with diabetes or those who may potentially develop diabetes.

Introduction

'I'm sorry, you have diabetes.'

These words are heard by over 100,000 patients annually in the UK alone, and carry a dark cloud of despair. Not only will the patient suffer from this seemingly relentless disease, but so will their family members. Diabetes is a troubling disease that, above all, is dangerous. Every year nearly 50,000 people die from diabetes, putting it in the top 10 leading causes of death in the UK.

Diabetes is a disease characterized by high levels of glucose or sugar in the bloodstream. In type 2 or 'age onset' diabetes, this high level of blood sugar is caused by the body's inability to utilize its own insulin to digest sugar. Medical researchers don't know why this happens and call it 'Syndrome X.' Caused by poor nutrition, weight gain and lack of exercise, type 2 diabetes is becoming one of the most prevalent health problems in the world. By early in the next century, it is expected to reach worldwide epidemic proportions.

A person with diabetes runs the risk of a host of other problems, including blindness, heart disease and strokes.

There are 1.4 million people, approximately 2 per cent of the population, who have diabetes in the UK, at a cost to treat of billions of pounds per year. It is believed that a further 450,000 adults in the UK have the disease and don't even know it.

How Chronic Illness Is Like Grief

Living day to day with a chronic illness like diabetes presents special challenges. Similar to the grieving process, you and the people around you may move through specific phases as you adjust to your illness and the demands it places on your body and life. Trying to avoid or ignore these feelings adds more stress, because it is only by moving through these phases that you can come to live a full, active life, despite diabetes.

The phases and the common response they illicit are:

- Shock and/or denial – 'I don't really have diabetes. My diabetes isn't that bad. This isn't happening to me. None of those complications will happen to me. I don't feel sick, I feel fine.'
- Anger – 'It's the doctor's fault. I'm not going to follow this treatment plan and you can't make me. I have to die of something.'
- Bargaining and/or negotiation – 'I'll just have this extra helping now and make up for it tomorrow. If I exercise more, this will all go away.'
- Depression and/or withdrawal – 'I give up. It's hopeless. This disease is controlling my life.'
- Acceptance – 'It's hard work to stick with the programme, but it's worth it.'

After the initial denial and anger in reaction to this diagnosis, most patients will begin some type of behaviour intervention, trying to undo years of nutritional, physiological and even psychological abuse.

Though this is not the most exciting and up-beat start to a book entitled *The Diabetes Cure*, the alarming statistics illustrate what those who get diagnosed with type 2 diabetes mellitus have to face.

Diabetes mellitus is a disease of glucose imbalance. The disease is well described in the medical literature, and most doctors treat it daily in their practices. Much of this standard treatment is ineffective and passive. The typical consultation with a doctor results in a review of the symptoms and a few tests. The patient is told he or she is either doing well or needs to continue to improve in one area or another. Seldom do doctors give their patients the education and information necessary to *cure* or reverse diabetes.

Type 2 diabetes is an eminently curable disease. I know, because I have cured it in myself.

My brush with diabetes began as an 18-year-old pre-medical student. Leaving high school at 6' tall and 12 stone, I entered the pre-medical programme at Montana State University in Bozeman. My goal was to become a well-rounded family doctor. I became 'well-rounded', but not in the way I had originally intended.

The strenuous exercise of high school gave way to an exercise routine that amounted to carrying 20 pounds of books to and from classes. Many days I sat in my dorm room studying while my friends engaged in serious exercise on the playing fields. It seemed as though everyone but me was outside playing football, Frisbee or chasing the opposite sex across the commons.

Along with strenuous study came strenuous eating. My interest in food stayed higher than my interest in exercise and fitness. I gained weight and ignored it or denied it. The person I saw in the mirror didn't look *that* heavy, I told myself. When my trousers became too tight I blamed it on the tumble dryer instead of those hot beef sandwiches I was eating every day.

Throughout my pre-med studies I worked at the local ambulance service as an Emergency Medical Technician, which provided a broad exposure to a full spectrum of disease and injury. Many patients who had congestive heart failure, renal failure, blindness and limbs that had been removed were actually patients afflicted with the final stages of uncontrolled diabetes mellitus. Little did I know, as a budding doctor, that these patients were portents of my own future.

After university I went on to medical school, though I didn't leave my habits behind. I was still eating like a pig and exercising like a snail when I walked in the doors of the University of Washington Medical School.

During Christmas break, I decided to return to the Emergency Room to see the nurses I had befriended. I hadn't seen them in four months and wanted to let them know that I was firmly ensconced in medical school. I was anxious for them to meet the new man I was becoming.

When I ran across the parking lot to the ER entrance on a bitter cold night, a small ache in my knee was barely noticeable as the large electric doors opened. The smell of alcohol and betadine reminded me I was home.

After a quick hello to the ER doctor on call and a high five to the janitor, one of my favourite nurses turned round. It had been four months since I had seen Jill and I stood there with my arms up, waiting for her to hug me. As she turned around, her bright smile quickly turned into a puzzled frown.

'What happened to you?' she demanded.

My delight disappeared. There was nowhere to hide. I caught sight of myself in one of the hallway windows and couldn't deny that I had changed. I was 3 stone heavier than when I'd left for med school. I had gained 10 pounds per month.

Jill continued to talk to me, but all I could hear in my mind was the classic quote from Socrates, 'Physician, heal thyself.'

The sad thing was that I *didn't* heal myself, not straight away at any rate. I continued to gain weight until I weighed 22 stone, and then I did something about it. I had been diagnosed with borderline diabetes and given a glucose-control drug to lessen the effects of glucose imbalance. On my own, I began to diet, exercise, take vitamins and minerals, and even experiment with the herb hydroxycitric acid which I would eventually prescribe to patients in my own practice. After several months of hard work and behavioural change, I cured my own diabetes. I have never taken a glucose-control medication again.

One of the ironies in my situation was that none of my professors at medical school ever told me I could cure diabetes. And, of course, no one researched the benefits of herbal remedies on type 2 diabetes. Instead, they were tightly focused on the art of giving medicine.

During the last eight years I have been a practising GP in Renton, Washington. I have had the privilege of caring for many patients afflicted with diabetes, borderline diabetes, advanced diabetes, complicated diabetes and patients who didn't know they had or would get diabetes. There are many people in this latter category.

The National Diabetic Prevention Trial in the US has uncovered the shocking fact that there are three times more people with diabetes or 'glucose intolerance' than have been detected. This is an alarming statistic, with dangerous potential implications if not aggressively combated. The complications of diabetes are devastating and affect nearly every organ of the body.

Diabetes can not only destroy a human life but it affects the people around the patient, who know that diabetes is a disease of lifestyle but maybe don't know how to change lifestyle habits to make their loved one healthier. Although many people are living with diabetes very effectively and managing their diabetes well, some are not. They are frustrated and searching for information that will allow them to launch a successful and sustained attack against their diabetes.

That is the goal of this book. By following a nine-point programme that combines the standard tools of diabetes control and prevention – exercise, diet, stress reduction and vitamin supplementation – with a mineral called chromium and a herb called 'HCA', you can prevent, control and even cure type 2 diabetes. I emphasize the prevention aspects of the nine-point programme because type 2 diabetes is one of the most preventable of all diseases. By following the nine-point programme *before* you have diabetes, you can reduce your chances of getting type 2 diabetes and a host of other 'diseases of lifestyle', including heart disease and strokes. The cliché 'prevention is better than cure' is repeated often because it is true.

The herbal portion of this nine-point programme, HCA, is technically known as *hydroxycitric acid*, a readily available substance found in the brindle berry, *Garcinia cambogia*, and other Asian fruits. HCA reduces cravings for sweets and enhances the effectiveness of the body's own insulin. Used for years as a herbal alternative to diet drugs, HCA allows the muscles of the body to absorb more glucose, the sugar in our bloodstream that is our body's main form of fuel. When this happens, a 'full' signal is sent to the brain, resulting in a reduced food intake.

The chromium taken with the HCA also helps with the absorption of glucose by the muscles. The utilization of glucose is important. Glucose that is not used by the muscles and other body tissues stays in the bloodstream and causes the dangerous rise in blood sugar that is known as diabetes. Elevated blood sugar causes damage to the organs and other body parts.

In addition to increasing the rate of glucose absorption, hydroxycitric acid also binds to left-over glucose in the bloodstream, slowing down the rate at which carbohydrate is converted to fat. The result is a natural approach to glucose control, without side-effects.

I discovered HCA while researching the scope of diet medications I could offer my weight loss patients, many of whom also have diabetes. I found that combining lifestyle change with HCA allowed patients with type 2 diabetes to go off medication almost immediately. For a natural approach to curing most cases of type 2 diabetes, HCA cannot be beat. Here are the reasons why:

- **There are no side-effects.** Unlike other diet drugs and the commonly used diabetes medications, there are no side-effects to using hydroxycitrate. Indeed, one of its main sources, the brindle berry, is commonly used as a flavouring in curries in many Asian countries. It is also used as a preservative and digestive aid. Researchers at Hoffman-Laroche, the pharmaceutical company that has conducted a large amount of HCA research, consider hydroxycitrate to be safer than citric acid in oranges and other citrus fruits.

- **HCA (especially in combination with chromium) makes the body more receptive to insulin.** People with type 2 diabetes usually have enough insulin in their bloodstream, but the muscles won't allow it to be used. When that happens, they overeat until there is so much glucose and insulin in their bloodstream that the body is forced to use it. It is the body's inability to use insulin that makes so many people with type 2 diabetes overweight.

- **HCA keeps glucose stores high.** A study in the *American Journal of Clinical Nutrition* shows that HCA keeps the body from converting carbohydrates into fat. It also slows the formation of cholesterol and triglycerides, two known causes of heart disease and strokes.

- **HCA keeps the person with diabetes from getting hungry.** By keeping glucose stores high, research has shown that this herb sends a message to the brain that the body is full and not hungry. This keeps the person with type 2 diabetes from gaining weight, as weight gain only makes diabetes worse.

I have seen changes in my own patients as they have participated in our comprehensive, multi-disciplinary diabetes and weight management programmes. They have learned through our Transitions Course the tools needed to make major lifestyle changes and keep these changes in effect.

The elements of the Transitions Course will be discussed throughout this book. We will help you understand the complicated inner working of your body and how various things you do affect its processes. Through knowledge and understanding follows compliance and control. The more you understand the information or knowledge we have about diabetes and how to prevent or cure it, the more compliant you will be with behavioural lifestyle changes and the more control you will exert over your life. Our intent is to help you to understand the knowledge we have about diabetes and to help you develop new skills to control or prevent diabetes.

My patients have proven to me that diabetes can be controlled and cured. They have proven to me that the devastating effects of diabetes can be prevented, diverted or reversed with much less effort than previously expected.

'To cure' is defined by the 1996 *Webster's II* dictionary as 'recovery from illness; a course of medical treatment for restoring health; a remedy; to rid of'. *The Diabetes Cure* is intended to help you help yourself and others to prevent or reverse the course of this devastating disease.

Vern S. Cherewatenko, MD, MEd

You may e-mail the author at: drvern@healthmax.net, or visit the *HealthMax* website for more detailed information on associated programmes and services at http://www.healthmax.net

CHAPTER ONE

Can Diabetes be Cured?

Diabetes is called the silent disease because it usually sneaks up on you without warning. It can start with feelings of overwhelming tiredness, or a need to eat even though you feel full to capacity. It sometimes shows up as blurry vision or as flashes of light before your eyes. You might have a headache that lingers long past the aspirin you have taken to get rid of it, or a cut or sore that just doesn't heal.

Like most people you will ignore these early signs by blaming them on stress or too little rest. After a while, however, the 'silent' disease becomes too noisy to ignore. Perhaps you lose several pounds without being on a diet, or you have a numbness or tingling in your hands that makes you fear a stroke or tumour. You know something is wrong that should be attended to by a doctor.

After hearing your symptoms, the doctor suspects that you may have diabetes. Still, a diagnosis can't be made until at least a Random Plasma Glucose Test is done. This test is the simplest way to detect diabetes because it measures the amount of blood glucose in your system. It can be done without fasting to see if diabetes is possibly the problem. The nurse draws a blood sample and you wait for an answer from the laboratory.

The results aren't good. Your doctor tells you that the

normal glucose count for this test is under 200 mg/dl. Your numbers, she explains, are over 300.

'Could there be a mistake?' you ask, feeling a wave of fear.

'Possibly,' says the doctor. 'The Random Plasma Glucose Test only tells us that you might have diabetes. The Fasting Plasma Glucose Test will tell us for sure if you have.'

For the next 8 hours you eat nothing before going back to the doctor for another blood test. 'I hope it's different this time,' says the nurse as she sticks the needle in your arm and draws a blood sample.

The next day you are back in the room with your doctor. She looks at the results of the blood test and begins to interpret them for you.

'In a patient without diabetes, the fasting glucose level will be less than 110 mg/dl,' she says, looking down at your results. 'You are substantially higher than that. I am afraid you have diabetes.'

A weight seems to land on your shoulders as the news settles in. Diabetes. You don't know how it happened, you don't know when it started, and chances are you don't really know exactly what it is. But you do know that you have it and that it can be bad if you don't take care of it. You think about other people in your family who have developed diabetes at your age, and you remember how they have wrestled with the problem. You might even know someone who has become seriously impaired from this silent but deadly disease.

'I know this isn't good news, but millions of people have diabetes,' says the doctor, trying to be reassuring.

'But I don't want diabetes,' you say. 'Can't you cure it?'

'Curing diabetes would take work on your part. Most patients aren't willing to do what it takes to cure diabetes. On the other hand, I can treat it with drugs. The drugs I can give you will level out your blood sugar, but they won't cure diabetes, only treat it. You will have to take these drugs for the rest of your life.'

'But I don't want to be dependent on drugs for the rest of my life,' you say. 'I am afraid of their side-effects.'

'All drugs have side-effects,' says your doctor. 'But if you don't take them, you will suffer from the complications of diabetes and those are much worse than the drugs used to treat it.'

The prospect of having to take drugs to survive is frightening. You trust your doctor and you know that the treatment she is offering is the same one being used for other people with diabetes. You don't want to go down that slow road of degeneration that so many other people with diabetes have followed. You want to fight against your disease, but you just don't know how.

'Doctor, I am willing to do what it takes to cure my diabetes,' you insist. 'But tell me, can diabetes be cured?'

Battling 'Syndrome X'

'Can Diabetes Be Cured?'

I have been asked this question many times by people who are facing a lifetime of diabetes medication. The simple answer I give to this question is a qualified 'Yes.' Type 2, or 'age onset' diabetes, is usually considered a disease of lifestyle. Unlike type 1 diabetes, which usually begins before the age of 20 when the pancreas fails to produce the insulin the body needs to process energy, type 2 diabetes generally begins after the age of 30 and is caused by weight gain and physical inactivity. The combination of these two factors makes your muscles resistant to insulin, the hormone secreted by the pancreas which converts blood sugar, or glucose, into energy. Just as petrol must be transported into the engine of a car for combustion to take place, glucose must be transported into your muscles for energy to be created. Insulin does that transporting.

FIGURE 1.1

The Pancreas

The pancreas is a large gland lying across the back of the abdominal wall. It has two secretions:

1. A digestive secretion poured into the duodenum.
2. A hormonal secretion passed into the bloodstream.

Capillaries drain the internal secretions of insulin and glucagon to the bloodstream to carry out their respective functions. It is the islets of Langerhans that produce the insulin in the body.

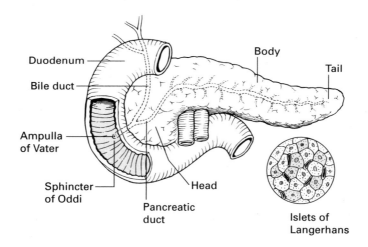

If your muscles are resistant to insulin, or your pancreas isn't creating enough insulin, glucose builds up in your blood, making it a thick sludge that is hard for the heart to pump round the body. Blood with too much glucose clogs arteries, causing heart attacks, strokes and a number of other very serious problems. Muscles that can't utilize large enough quantities of glucose waste away, while nerves that don't get glucose die.

The cause of insulin resistance isn't known and is called 'Syndrome X', a name coined by Dr Gerald Reaven of Stanford University in California.

In general terms, with Syndrome X the cells don't let insulin do its job, which is to 'open up' the cells and let glucose enter and be used as energy. When this happens, the levels of glucose in the blood become higher and higher. Excess body fat and too little exercise combine to create Syndrome X. Sometimes the pancreas must work overtime to produce insulin. An overworked pancreas can become damaged, leading to an organ that fails to produce very much insulin at all. If that occurs you might have to take artificial insulin.

Syndrome X is a double-edged sword. Not only is it the main cause of type 2 diabetes, it also causes you to gain weight, which makes the disease worse.

Weight gain takes place when the lack of glucose in your cells blocks your satiety response to food – that feeling of being full and no longer wanting to eat. The feeling of satiety is related to the amount of glucose contained in your muscles and brain. If you are insulin resistant, glucose takes longer to get into your brain's hunger centres, so you keep eating until it finally does. In the meantime, glucose reaches higher and higher levels in the bloodstream, which leads to the many symptoms and complications of diabetes.

Although some prescription medications have been very effective in normalizing glucose levels, they can also lead to sharp drops in glucose levels which force you to eat larger quantities of food. When that happens, you gain more weight and your condition becomes worse.

After relying for years on prescription drugs for the treatment of type 2 diabetes, your disease nevertheless generally becomes progressively more severe.

Waging War on Diabetes

Although allopathic medicine has done a good job of extending the lives of patients with diabetes, it has not done a very good job of curing or preventing this disease. Many references in medical books declare that 'there is no cure for diabetes at this time.' I attended one of the most prestigious medical schools in the country and I never heard one of my professors say that diabetes could be cured.

The fact is that it *can* be cured. By taking the multi-disciplinary approach outlined in this book, myself and other doctors have been able to reduce the dependency patients have on prescription diabetes drugs, or enable these patients to eliminate them altogether. In my programme I use a common weight loss herb known as hydroxycitric acid, or HCA. A readily available substance found in the brindle berry and other fruits, HCA reduces cravings for sweets and enhances the effectiveness of the body's own insulin. HCA has been used for years as a herbal alternative to diet drugs.

However, unlike these controversial diet drugs which work by altering brain chemistry, HCA allows the muscles of the body to absorb more glucose. When this happens, a 'full' signal is sent to the brain, resulting in a reduced food intake.

In addition to increasing the rate of glucose absorption, hydroxycitric acid also binds to left-over glucose in the bloodstream, slowing down the rate at which glucose is converted to fat. The result is a natural approach to glucose control, without dangerous side-effects.

A Natural Approach

I discovered HCA while researching the scope of diet medications I could offer my patients. As a bariatric physician – one who specializes in weight loss – I have treated more than 5,000 people for obesity. Many of these patients have type 2 diabetes and have taken diabetes medication for years.

When they come to my clinic, I start them on a comprehensive programme that involves regular exercise, sensible eating, vitamin and mineral supplements and medication to curb their appetite. I have had the opportunity to see the effects on type 2 diabetes of virtually every diet drug ever marketed. I have found that a combination of lifestyle change with any diet drug will allow a person with type 2 diabetes to eliminate their medication or greatly reduce the amount they need to take, almost immediately. For a natural approach to curing or lessening the effects of type 2 diabetes, HCA cannot be beat.

Why Not Diet Drugs?

The notion that diet drugs — HCA or any other type — have an effect on type 2 diabetes might be hard to believe. Aren't these drugs aimed just at making you lose weight?

The answer is 'yes.' But because of the weight they make you lose, they have other effects on the human body as well. Several studies have shown the positive effects of diet drugs on type 2 diabetes. For example, a 1995 double-blind study done in New Zealand found that insulin sensitivity was dramatically increased in patients taking dexfenfluramine, a popular diet drug. This insulin sensitivity increased after only one week of therapy and before any weight loss.

Other studies have had the same results as the one done in New Zealand. Repeated controlled studies have shown 'improved glycaemic control' among people with type 2 diabetes through the use of diet drugs. Many diet doctors have noted anecdotally that many of their patients with diabetes are able to stop or greatly reduce their use of type 2 medication within a week or two of starting diet drugs.

Of the thousands of patients treated in my clinics, many people with diabetes have gone medication-free as a result of taking diet drugs. In addition, insulin-dependent patients have been able to reduce the amount of insulin they take by as much as a third, just by taking diet drugs of all kinds, including HCA.

Although it doesn't work as quickly as prescription diet drugs like Fen Phen and Redux (both of which are now in any case banned), HCA doesn't have any of the side-effects of these medications. With Fen Phen and Redux, pulmonary hypertension (a critical form of high blood pressure) was a deadly and hard-to-detect side-effect. Another possible side-effect of prescription diet drugs is serious fluctuations in neurotransmitter levels in the brain, which could have long-term health consequences.

Medical science does not yet know the full effects of many diet drugs. Recent research has shown that these diet drugs may carry a substantial health risk. A Mayo Clinic study, for example, linked the diet drug Fen Phen to heart disease in women, resulting in a warning against its use from the doctors who conducted the research. An article in *Hospital Practice* discussed the positive effects of diet drugs on type 2 diabetes, but then issued a warning to doctors who might consider pre-scribing them to patients: 'Given the risks and caveats associated with anorectic drug therapy,' the authors wrote, 'one might wonder whether the potential benefits are worth the effort.'

HCA, on the other hand, has no side-effects. Ten years' worth of research — both animal and human — shows that HCA reduces appetite by means of increased insulin sensitivity. HCA also slows the process of converting fat into triglycerides, a form of blood fat that is linked to a higher incidence of heart disease and strokes. Here is more research showing the effect of HCA on the causes of type 2 diabetes:

- In an eight-week double-blind study, 50 obese volunteers were placed on a low-fat diet. Twenty-nine of the patients were given HCA before each meal, while the rest were given a placebo. After eight weeks, the HCA group lost an average of 11 pounds per person, while the placebo group lost only 4 pounds per person. The reason given for the added weight loss was greater appetite suppression caused by increased insulin sensitivity.
- Rats were fed a high-fat diet supplemented with HCA and were compared with rats that were fed a high-fat diet that was not

supplemented with HCA. Each group was fed its respective diet for 15 days. The HCA group ate less, lost weight and had lower levels of serum triglycerides than the other group. Much of the effect was attributed to the fact that increased insulin sensitivity reduces appetite.

- In two human trials published in *The Bariatrician*, two groups of patients were asked to assess their level of energy after one group was given HCA and another was given a placebo. The group given HCA consistently reported more energy than the group given the placebo. All were then blood tested. The HCA group had far lower levels of blood sugar than the placebo groups, indicating that the blood sugar had been absorbed into their system.

Case Study – Jim

There are a number of studies showing the effectiveness of HCA in weight loss, reduction of blood serum cholesterol and triglycerides and — most important for the person with diabetes — its ability to increase insulin sensitivity. But there is no greater test of a drug than seeing it in action with a patient.

Typical of the cases I have seen is a man I'll call Jim, an overweight 43-year-old with a fasting glucose level of 176 mg/dl, which put him well into the blood sugar range for type 2 diabetes.

Jim's diabetes was not his only concern. Other tests found that he was also at risk for cardiovascular disease. His cholesterol level was 356 – 156 points higher than the ideal recommended by the American Heart Association. Triglycerides were a whopping 990, more than 700 points higher than the ideal. Numbers like these guaranteed that Jim's diabetes would worsen and cardiovascular disease would develop.

I put Jim on 1,000 milligrams of HCA per day, a low-fat diet and an exercise routine that had him walking between 30 and 60 minutes a day.

In less than two weeks he was tested again. His glucose level had dropped to 111, well within normal. Also, total cholesterol was reduced to 262. The most amazing reduction was his triglyceride level, which fell by almost 600 points to 396.

During the next two months Jim's numbers continued to drop. As of his most recent check-up, his glucose level was 85. His cholesterol level continued to drop to 185, and his triglyceride level to 117. With numbers like this, clearly not only Jim's diabetes has reversed. Jim's blood composition is now in a range that can actually *reverse* the cardiovascular disease that might otherwise have occurred as a result of years of high cholesterol and borderline diabetes.

A Rising Problem

The most startling aspect of this case study is not that Jim was able to cure his diabetes. For me, it was his age — 43. Since 1958, the number of people who have been diagnosed with diabetes has increased sixfold. According to the US Centers for Disease Control and Prevention, another six million people in the US are unaware that they have the disease. Other studies place that figure even higher, estimating that half of adults with diabetes are undiagnosed.

More people worldwide have diabetes than ever before. A recent estimate puts the global number of people with type 2 diabetes at 120 million. In addition, more of those people with diabetes are younger than ever before. More than two-thirds of the known type 2 diabetes sufferers in the UK are 'baby boomers' born between 1946 and 1964. A growing number are children, a frightening scenario since there are no studies to guide us on how to provide safe treatment. Since type 2 diabetes in children is such a recent phenomenon, there have not yet been any studies.

What is causing this sudden and catastrophic rise in type 2 diabetes? The answer is poor diet and lack of exercise. The high-fat diet of the West has been adopted worldwide, with fatty fast foods replacing the rice- and vegetable-based diets that used to dominate in many countries. Along with diets higher in fat has come a more sedentary lifestyle. Most people in the world don't do as much physical work or play as they used to. This lack of proper diet and exercise has led to obesity, which is present in 95 per cent of people with type 2 diabetes.

The number of people with diabetes who don't know they have the disease has alarmed the medical community to such an extent that massive screenings have been suggested by some researchers, to stave off 'the beginning of an epidemic'.

History of the Mystery

It has only been in this century that the causes and treatment of diabetes have been known. Before that, it was a mysterious disease, the cause of which was only a guess.

The earliest known record of diabetes, written on papyrus by Hesy-Ra, a 3rd Dynasty Egyptian physician, discussed frequent urination, or 'polyuria', as a symptom. During the first century AD, Arateus described diabetes as 'the melting down of flesh and limbs into urine', while Galen of Pergamum, a Greek physician, felt diabetes was a form of kidney failure.

The diagnosis of diabetes was frequently confirmed by 'water tasters' who determined the sweetness of sufferers' urine by drinking it. The term *mellitus* is derived from the Latin word for honey. Diabetes mellitus was used to describe the disease of sweet urine.

Early in the 19th century, urine testing became more scientific when the first chemical tests of urine were developed to measure levels of sugar in the urine. The French physician Priorry recommended high doses of sugar as a treatment during the late 1850s. Twenty years later Bouchardat, another French physician, identified that the excess glucose in the urine went away in his diabetes patients during the Franco-Prussian War food-rationing programme.

Also in the 19th century, Claude Bernard researched glucose metabolism in the liver and studied the physiology of the pancreas. This began the increased scientific study of the relationship between the pancreas and glucose metabolism. Later that century, Italian physician Catoni would lock patients up to ensure they'd stick to diet modifications he would prescribe.

In 1869 a devoted German medical student revealed his research regarding a two-cell system in the pancreas. One set of cells was felt to secrete digestive juices; the other set had an unknown function. This student's name was Paul Langerhans, and the two cell sets have come to be known as the 'Islets of Langerhans'. It is these cells that produce insulin. Twenty years after Langerhans described the two-cell system, a pair of Austrian researchers called Minkowski and von Mering surgically removed the pancreas from a dog for examination.

The first 'fad diets' came on the scene in the early 1900s. These included eating an abundance of single items such as potatoes, rice, oatmeal and milk. George Zuelzer, from Germany, gave the first injections of 'pancreatic extract' in 1908, which resulted in many severe side-effects.

During the latter part of 1920 through 1922, a Dr Banting researched, developed and tested pancreatic extracts for their effectiveness on diabetes.

On May 21 1922, James Havens became the first American to receive insulin successfully. When Eli Lilly and Company began the manufacture of insulin, the treatment of diabetes became more widespread. Banting and Macleod received the Nobel Prize for Medicine in 1923, with recognition of Drs Best and Collip for their pioneering work on insulin development.

Further research in the 1940s gave insight into the relationship of diabetes to other diseases, namely kidney disease and eye degeneration. The insulin syringe was developed in 1944, and the first oral medications for diabetes were seen in 1955 to help patients lower blood sugar levels. Type 1 diabetes and type 2 diabetes were recognized as separate syndromes in

1959, with type 1 dependent on insulin treatment, and type 2 not dependent on insulin.

Patient education expanded in the 1960s with recommendations to test urine at home to increase glucose control. Public awareness is beginning to grow as results of the Diabetes Control and Complications Trial (DCCT) are evaluated. Research has determined that there is a far greater number of undiagnosed people with diabetes, and efforts are underway to identify them.

Why Alternative Therapy?

For years, the notion of using alternative therapy to treat type 2 diabetes has been surpassed by the use of medications. It seems as though every year has at least one new treatment for diabetes. One of the most recent, Rezulin, was pulled from the European market when patients died and studies showed that it had side-effects that were dangerous to the liver.

Many people have relied upon drugs and little more to treat their disease. They have mistakenly thought that only glucose-control drugs can bring them back to normal. Don't get me wrong, the glucose-control drugs discussed in the next chapter are great, but they have side-effects which can include a worsening of the disease. Although they are prescribed for almost everyone with type 2 diabetes, they are not necessarily for everyone.

More and more people with diabetes are looking for an approach to their disease that has no side-effects and gives them the possibility of curing diabetes with a natural method that actually makes them healthier.

In the last few years I have seen an increase in the demand for alternative treatments as patients realize that many drugs just conceal the disease process and don't really slow or stop it. These patients are the ones who are willing to do more than simply take a pill. Pills alone won't do it. Even the most powerful of the diabetes drugs can't stop diabetes if the patient will not implement the lifestyle changes necessary to control his or her glucose levels.

That is why my 'Diabetes Cure' requires more than just a few tablets of HCA per day. The prescription I am writing with this book calls for a multi-disciplinary approach to glucose control, involving:

- daily HCA
- thirty minutes to an hour a day of enjoyable exercise
- monitoring the type and amount of food you eat
- daily vitamin and mineral supplementation – especially of chromium – to add punch to your body's metabolism and defences
- daily stress reduction that gives you time for yourself and gives your body time to regroup.

Each of these aspects of the Diabetes Cure will be covered in detail in the pages ahead. Much of this is information that should have been given to you clearly and carefully when you first found out about your disease. Chances are it wasn't. Doctors today have less time for patients because managed care has forced them to spend more time on business. Studies have shown that the average doctor is spending seven minutes completing paperwork for every one minute he or she spends with a patient. This 'patient-to-paperwork' ratio has had a direct effect upon patient education. With little time to spare, doctors are prescribing diabetes medications and sending their patients out the door when they should be helping them make necessary lifestyle changes.

Many in the medical community are concerned about this reliance on medication at the cost of patient education. In *Clinical Diabetes*, a medical journal published by the American Diabetic Association, editor Alan J Garber, MD, PhD expressed the concern of those of us who believe in a more holistic approach to this disease.

Managed care organizations until recently generally have not espoused the commitment to patient education required to make diet and exercise a meaningful prescription. Physicians have all too willingly participated in the prescription-writing frenzy that is often necessary to control Type 2 diabetes ...

Thus, the recent proliferation of effective oral therapies is both a blessing and a curse. It is a blessing for patients who are actively attempting to care for their diabetes and for the physicians who participate in such efforts. But it may be a curse for patients who can delude themselves and their physicians that such agents are effective even in the absence of diet and exercise, points which have not been demonstrated in prospective, randomized, clinical trials.

It's Up To You!

At its very heart, the Diabetes Cure is about taking charge of your own disease. By educating yourself with the information in this book, you can slow, stop or even cure your type 2 diabetes.

Most people respond to a diagnosis of diabetes by becoming depressed. Not only do they have a fear of the unknown, but they also fear 'making a mistake' that will raise their glucose levels and cause damage. There is, of course, reason to fear diabetes. It can be a relentless disease that chips away at your weak spots and causes complications like heart disease, vision problems and strokes. On the other hand, knowledge and understanding of this disease can put control back into your life.

Of course taking the holistic road and trying to cure a disease is harder than just managing it. When patients express doubt about their ability to change their lifestyle rather than rely on the false security of medication, I ask them to answer three questions:

- If not this, what?
- If not you, who?

- If not now, when?

Good luck as you begin the transformation that will give you control over your disease.

CHAPTER TWO

Fighting a Winning Battle

A patient came to me with the frustration felt by many people with diabetes. Phil had been diagnosed more than four years earlier with type 2 diabetes. His doctor had given him a drug that would increase the amount of insulin produced by his own pancreas. Although he had plenty of insulin circulating in his blood already, his cells had become resistant to insulin. By increasing its amount, the cells would be forced to open up to insulin and let the glucose in to be used as energy.

This medication worked fine for about a year. Then Phil began to notice that he had gained a considerable amount of weight. Concerned about the dangers of gaining weight, he asked his doctor for help and was given information about diet and exercise, the first time he remembered hearing about either one in relation to his disease.

He was fairly diligent in following the advice, although he admitted that he did not exercise every day and frequently 'fell off the wagon' when it came to eating right. Still he was gaining weight and not feeling so good, either. Some of his symptoms were returning. He would have periods of binge eating that he just could not avoid.

Over the course of three years, he had gained more than 2 stone. This shocked him. Although he had always been a little on the heavy side, he had never gained weight as rapidly as when he started his medication for diabetes. When his doctor increased the amount of medication he was taking in hopes of getting rid of some of his symptoms, Phil began to feel even worse. At times he felt literally driven to the refrigerator by feelings of hypoglycaemia, or low blood sugar. He was also experiencing extraordinary thirst.

The return of Phil's diabetes symptoms led his doctor to increase his dosage of medication. If his symptoms worsened, Phil's doctor said he might add *another* medication to boost the effectiveness of the one he was already getting. If his problems continued to worsen, Phil was told he might be forced to add an injection of insulin to his daily medication regimen.

Phil's concern that his condition was not improving led him to me. I had treated a friend of his for obesity, which also led to curing his borderline diabetes, the early stages before type 2 diabetes is diagnosed.

I had him follow the Diabetes Cure to the letter, with particular emphasis on exercise and stress reduction. By administering the 11—11 Scale contained in Chapter 7 (page 153), I could tell that he needed relief from several stress factors in his life. I did not take him off his medication, since a person with diabetes should never stop medication 'cold turkey'. Rather, I started him on daily doses of HCA and chromium and had him watch his glucose carefully.

Within three months Phil had lost close to 10 pounds and was feeling free of symptoms. I was gradually able to wean him from his medication until he was taking about half the amount he had been when he first came to see me.

Phil has continued to improve, and the longer he follows the steps in the Diabetes Cure the more he will improve. He now realizes two very important things about diabetes:

1   It is a proactive disease, one which you must fight constantly on every front.

2   Diabetes drugs do not cure the disease and were never intended to. Taken alone, they can only slow the progression of this relentless disease.

Symptoms, Not Cause

Historically, diabetes treatment has been aimed at treating the symptoms of the disease and not its cause. That means the goal of most diabetes drugs is to prevent the problems diabetes causes, like excessive thirst, hunger, urination, irritability, weakness and fatigue. They do relieve these symptoms, but they do it without making the body more receptive to insulin, which is the cause of 95 per cent of all cases of type 2 diabetes.

This is the best that most diabetes drugs can do. Most of these drugs work by increasing the amount of insulin in your blood, slowing the release of glucose by your liver, or slowing the absorption of food by your intestines. Since they do not increase insulin sensitivity, these drugs alone will not win the war. They will only help you fight a losing battle.

People with diabetes who rely heavily on drugs generally do not have good glucose control. Every time glucose moves into the danger zone (above 200 mg/dl), more damage is done to the body, and complications are more likely to arise.

This was proven in the Diabetes Control and Complications Trial (DCCT), a 10-year US National Institutes of Health study of glucose control in people with diabetes, mentioned earlier. Their study showed that hyperglycaemia, or high blood sugar, leads to nerve and vascular damage that can increase the risk of eye, nerve and kidney disease. The good news is that careful blood glucose control through an intensive regimen of diet and exercise reduces the risk of complications by 50 to 75 per cent.

The Diabetes Drugs

Doctors have five different 'classes' of prescription drugs to treat type 2 Diabetes. All of these have side-effects and none of them can be relied upon alone. They all require a multi-disciplinary approach to fighting the disease of diabetes. Without incorporating at least diet and exercise as part of your treatment, you are fighting a losing battle against a disease that never rests, no matter how powerful the drug is that your doctor has prescribed.

Sulphonylureas

This long-standing class of drugs has been the mainstay of the oral medications for many years. **Minodiab** and **Diabinese** are the names of some of the more popular sulphonylureas.

These drugs stimulate the pancreas, increasing the amount of insulin that is secreted into the bloodstream. This added insulin forces the glucose into the muscles like a chemical battering ram.

The problem with sulphonylureas is that they don't improve your insulin sensitivity, rather they increase the amount of insulin in the blood. This solves the short-term problem by lowering your blood glucose levels, but it creates long-term problems by encouraging weight gain. This drug increases insulin, but some type 2 patients have insulin levels *five times* above normal levels already.

The process of increasing insulin levels leaves many people with diabetes hypoglycaemic. As a result, the sudden drop in blood sugar that this class of drug causes compels them to eat. People with hypoglycaemia get such symptoms of low blood sugar as mental confusion, dizziness and even loss of consciousness.

The result of hypoglycaemia is overeating, which leads to weight gain, which leads to increasingly poor insulin sensitivity. This is precisely opposite to the effect we want to see.

People with diabetes should lose those unnecessary pounds, not accumulate more!

Sulphonylureas are known to increase cholesterol and triglyceride levels, both of which are linked to heart disease.

There are currently nine sulphonylureas to choose from. **Glipizide**, sold in the UK under the brand names Minodiab and Glibenese, is one of the improved or 'second-generation' drugs.

Biguanides

Developed in 1994, biguanides approach glucose control from the opposite direction than sulphonylureas. Rather than increase the amount of insulin coming from the pancreas, they slow the amount of glucose entering the blood from the liver and the small intestines where food is digested.

The biguanides are an improvement over the sulphonylureas, because they do not cause weight gain or increase the risk of heart disease. They also have been shown to increase the levels of HDL or 'good' cholesterol while lowering triglyceride levels.

The benefits of this medication have their cost. People taking biguanides report bloating, nausea, cramps, abdominal fullness and diarrhoea. People with heart, liver or kidney problems are discouraged from taking this medication.

Typical of the biguanides is **metformin**.

Alpha-glucosidase Inhibitors

The alpha-glucosidase inhibitors such as **acarbose** slow the breakdown of starch in the small intestines by interfering with the alpha-glucosidase enzyme. This enzyme is responsible for breaking down food into glucose. When the enzyme's action is blocked, a rise in blood glucose levels is prevented.

Although effective in blocking glucose, it can cause some unfriendly side-effects. Excessive wind and flatulence, accompanied by diarrhoea and abdominal pain, make it a potential source of social accidents. These symptoms make it a drug that should be avoided by people with irritable bowel disease, ulcers and colitis.

Elevated liver function can also occur with acarbose, making it dangerous for people with liver problems.

Thiszolidinediones

The newest class of diabetes drug is perhaps the most controversial. Brand named **Rezulin**, this drug was the first on the market to help the body make better use of its insulin by stimulating receptor sites in tissue, the 'doorways' through which glucose passes from the blood. At the time of writing this drug is unavailable in the UK.

Some studies have tested the use of rezulin with sulphonylureas. Although it lowered the average glucose levels, patients in one study still gained 5.8 to 13.1 pounds.

Rezulin is not a magic bullet, however, and is not without side-effects. Like other medications used to treat type 2 diabetes, it has serious risks. Cases of liver problems have been reported, including fatality, which have led to the drug being banned in Europe. For this reason, patients are required to undergo regular liver function tests. Headaches have also been reported, along with infections and general pain.

Insulins – New and Old

Insulin is always used with type 1 diabetes, since people with that form of the disease do not produce insulin of their own. With type 2 diabetes, insulin is used as a last resort.

About 29 per cent of people with type 2 diabetes are eventually treated with insulin. Treatment of type 2 diabetes with insulin could usually be avoided with adherence to good diet and exercise routines. Although diabetes educators do a very good job in educating patients who need insulin, it is often information that arrives too late. By the time many patients find out they need insulin, they are often less motivated to make the changes called for to reverse their condition. They have given up, and are willing accept that their diabetes will be fought by medication only.

Insulin, available since 1922, is designed either to act quickly or to act several hours after administration. Whether short- or long-acting, insulin tells the liver to stop producing glucose and to start using it. In addition, insulin tells the muscles to suck in glucose and to start burning it for fuel. Insulin can also tell the fat cells to take in glucose, but this occurs to a lesser degree.

Insulin dosage and timing have profound effects on the blood glucose levels. Too much, and the blood glucose level falls, leading to hypoglycaemia. Too little insulin, and there is not enough to drive the glucose into the cells that need it. Insulin can be used by people with type 2 diabetes with or without some of the oral medications. Usual doses of insulin are given two times a day, morning and evening, to try to keep the glucose levels at or near normal range.

The delicate balance between such factors as food intake, energy expenditure through exercise, medication dosages, illness and stress all interact to raise or lower the levels of glucose in the blood. The more closely the blood glucose levels stay to the normal range, the fewer systemic effects diabetes will have.

As wonderful as insulin is, it has side-effects like all diabetes medications. Hypoglycaemia, weight gain and rapid occurrence of blood vessel disease are all potential side-effects of insulin therapy. These side-effects make it necessary to have close supervision by a health care provider trained in diabetes when undergoing insulin treatment.

So What Is HCA?

The plant kingdom has always been a medicine chest for humans. For millions of years nature was the only place we could go for medicine. Even today, more than 30 per cent of our medications are derived from plants. Among the drugs that have come from the plant kingdom are aspirin, which was first found to be contained in willow bark as far back as ancient Greece. Digitalis, used to treat cardiac arrhythmias, was refined from the foxglove plant.

It is estimated that approximately 80 per cent of the world's population uses herbal medicines, many because they prefer a more gentle and natural approach to treatment.

Among the thousands of plants used effectively is a reddish plant the size of a small orange with the botanical name *Garcinia cambogia*. When this plant is dried and ground into powder, it has uses for many different diseases. In India, for instance, *Garcinia* is sometimes used as a topical application for infections. It also has use as a 'chewing stick' to keep teeth and gums healthy. In some countries it is used as an aphrodisiac and even as a treatment for parasites and liver problems. In China, herbalists give *Garcinia* to patients with type 2 diabetes and heart problems.

It was the use of *Garcinia cambogia* in traditional treatments that led to an interest in its essence by researchers. Chemical researchers knew that it worked as a treatment for a variety of conditions, but they didn't know why. In 1965, two researchers found that the active ingredient in this herbal remedy was hydroxycitric acid (HCA), and research on its effects began. What they found with HCA was a chemical compound that did more than expected.

A chemist at Brandeis University, J A Watson, PhD, found by surprise that hydroxycitrate would trick cells into not synthesizing fat. If cells could not synthesize fat, then they would not store fat, which meant there would be less build-up of fat in the liver and on the body.

Several rat studies were undertaken in the 1970s to confirm Watson's findings. Researchers at Brandeis University and at Hoffman—La Roche Laboratories used rats to demonstrate that HCA reduced the synthesis of fat in living animals. They then began to speculate that HCA might have uses as a weight-control drug.

Researchers Ann Sullivan and Joseph Triscari fed HCA to test animals in their regular diet and found that they did not gain weight as fast as another group of test animals fed the same amount of food. Further examination of the test animals led them to the realization that HCA blocks fat accumulation.

In simple terms, HCA blocks fat synthesis by interfering with part of the 'Krebs cycle', the chemical process by which our cells convert energy into fuel. When food is eaten in greater quantities than our bodies need on any given day, the extra energy is converted into fat. HCA disrupts the cycle that converts this energy into fat, causing the excess fat to be excreted from our bodies instead of absorbed on our thighs and other places.

Further Discovery

Further animal research led to another discovery: rats that were fed HCA became satiated or 'full' on less food than the control group. They ate less food because the HCA made them less hungry. Further study of the animals told the researchers why satiety was reached so quickly. The rate of *gluconeogenesis* – food converted to glucose by the liver – was nearly doubled in the animals fed HCA. When the amount of glucose in the liver is high, it signals the brain through the vagus nerve that it needs no more food, and the person stops eating.

Additional research on animals found that HCA probably increases metabolic rate and speeds the burning of fat by the liver.

All of these factors – lower synthesis of fat, satiation or becoming 'full' on less food, and speeding the burning of fat by the liver – can have an extremely positive effect upon people with type 2 diabetes.

Further animal research led to another finding that might be the most important factor of all in the control of type 2 diabetes: HCA greatly reduced the amount of fatty acids in the bloodstream. In some studies, animals who were fed a combination of HCA and a high-fat diet had one-third the levels of fatty acids as rats fed the high-fat diet alone.

Lower blood fats is of particular importance to people with type 2 diabetes, since some researchers speculate that elevated levels of free fatty acids like triglycerides and LDL cholesterol can lead to insulin insensitivity, that 'Syndrome X' referred to earlier which causes the disease.

Research into HCA has been halted by Hoffman—LaRoche because the drug company could not patent an extract of a natural product.

Other Research

Despite the lack of funding from drug companies, research into the effects of HCA has continued. Its effects on body weight, blood lipids, insulin sensitivity and satiety – all of which are important factors in the control and cure of type 2 diabetes – have continued to be evaluated. When it comes to altering the factors that lead to type 2 diabetes, HCA passes the test.

An eight week study examined 54 people who had the risk factors for diabetes (overweight, inactive, poor blood lipid profile). Twenty-two of these subjects took HCA before each meal, while the remainder were given a placebo. After eight weeks, the study group had lost an average of 11.1 pounds per person while the placebo group lost 4.2 pounds per person.

The HCA group had an easier time making recommended lifestyle changes, including eating a diet low in fat and following a daily regimen of sensible exercise, probably because the HCA left them feeling 'full' after eating less food.

The HCA group also had a marked decrease in serum cholesterol and triglycerides. They also reported a higher level of energy, possibly caused by an increase in insulin sensitivity.

Three medications were tested on 128 patients over an eight-week period. Two of the groups were given two different diet drugs that did not contain HCA. The third group was given a medication containing HCA. The 29 subjects in the HCA group lost an average of 11.48 pounds per person, while the other two groups lost 9.52 and 6.65 pounds per person respectively.

Hunger levels were significantly lower in the HCA group, as were lipid levels. Energy levels were higher, however.

Volunteers were given HCA for eight weeks to study its effects on obesity, a major risk factor of type 2 diabetes. The study group consisted of 60 women and 15 men who weighed from $9\frac{1}{2}$ to 18 stone and varied in age from 21 to 65. The 250-mg daily dosage of HCA, taken 30 minutes before each meal, was supplemented with 100 mcg of chromium picolinate. Elevated blood lipids affected 74 per cent of the study group, and 53 per cent were taking prescription drugs for medical conditions related to elevated lipid levels and obesity, including diabetes medication. All of the patients were taught how to alter their dietary habits and follow a sensible exercise programme that included daily exercise.

The 42 subjects who completed the study lost an average of 10.8 pounds per person. One-third of the subjects had a significant lowering of their blood lipids, including cholesterol and triglycerides levels. In eight of the cases where diabetes medication was involved, glucose levels were reduced significantly.

A study published in June 1998 revealed all natural weight-loss supplements provided valid results. Supplements containing HCA and chromium were given before meals with water. A non-restricted diet regimen was used, which meant that the study subjects could continue with their usual diet.

Overall weight losses ranged from 10 to 26 pounds during the 10-week study period. Measurement of inches lost ranged from 4.25 to 26.25 inches during the study period.

> All participants reported increased energy levels. The study concluded that HCA is a viable form of weight-management intervention.
>
> With the direct relationship of diabetes and excess weight, this is yet another confirming look at the benefits of HCA as a natural supplement to improve glucose control.

Even with the many studies showing the benefits of HCA in weight reduction, glucose management and overall sense of well-being, a large, placebo-controlled, double blind study would be the best way to document these benefits scientifically. A study of this type is underway in several cities and multiple clinics across the United States.

According to Anthony Conte, MD, director of the LifeStyle Medical Center in Hilton Head Island, South Carolina, HCA has proven to be an excellent therapeutic approach to the treatment of obesity and weight-related disorders like type 2 diabetes. Conte has researched HCA extensively, and has published several articles using HCA as the major component in various supplements designed to help people lose weight and control their glucose levels. Conte states that during the last eight years since he completed a double-blind study on the effectiveness and safety of HCA there have been substantial medical improvements in those patients taking HCA.

The medical effects of HCA in his study include improved blood sugar levels, as needed for control of diabetes, but also a lowering of cholesterol levels, reduction in cravings for sweets, increased lean body mass, increased fat loss, improved energy levels and most certainly a decrease in overall weight.

Conte has been the primary investigator in several consumer studies and has seen the same positive effects in his personal patients who see him at the LifeStyle Medical Center. Conte feels that HCA, combined with lifestyle changes, can lead to a reversal of the glucose levels in patients with diabetes.

'We have noticed, aside from fat loss, an increased energy level, reduction of appetite, reduction of cravings for sweets, and a definite reduction in blood sugar levels, cholesterol and triglycerides,' says Conte. 'All of these factors combined can contribute to the reversal of type 2 diabetes.'

Here is a case history of one of Conte's patients who benefited within eight weeks from the HCA (containing supplemental chromium) and elements of the Diabetes Cure:

Case Study – Anne Marie

Anne Marie is a pleasant, middle-aged mother, 51 years old, who weighed in at 18 stone when she sought the help of Conte and his staff. At only 5'4", her excess weight had taken its toll in many ways. She had struggled with obesity nearly all her life, and had been on and off diets, trying everything under the sun to get her weight down.

Initial blood work revealed type 2 diabetes with a blood glucose level of 172. Her cholesterol was abnormally high at 321, and her triglycerides, or blood fats, were also elevated to 276. Conte went to work. Educating Anne Marie on the adverse effects of high glucose levels and the mechanisms behind the HCA, or brindle berry extract. She began to understand the interrelationship between diabetes and obesity.

Within eight weeks Anne Marie was on her way to reversing a long-standing trend. Her glucose level fell to 128 in those eight weeks of HCA use, her cholesterol had decreased to 217 and the triglycerides were down to 226. She had also lost nearly 20 pounds, and said she felt healthier as a result of being lighter.

With such clear success, she began to focus her priorities on what she needed to do to avoid a life of diabetes and obesity. She scheduled time for daily exercise. By doing that, she realized that so many of her commitments were a waste of time. She eliminated many of the 'unnecessary' things she had been involved in and scheduled more time for relaxation and stress reduction.

Her status has continued to improve, mainly because she started to take her disease seriously.

The One-Two Punch

Although HCA alone has a powerful effect on insulin sensitivity, I add chromium to give the Diabetes Cure that added punch. Working together, HCA and chromium can increase insulin sensitivity without added side-effects.

Studies examining chromium alone have shown that it:

- increases insulin sensitivity by stimulating glucose and amino acid uptake by cells
- increases muscle mass in exercising humans, which is important for people with diabetes since muscle 'soaks up' and burns glucose
- reduces the amount of diabetes medication required when taken in combination with other diabetes medications. In an Israeli study, 47 per cent of the patients who took chromium were able to reduce their levels of insulin or other medication.
- reduces cholesterol and triglyceride levels in people with diabetes, two factors that help create complications of diabetes.

The most exciting chromium study of all was conducted by the US Department of Agriculture's Nutritional Research Center. They gave a total daily dose of 1,000 micrograms to a large population of people with type 2 diabetes in China. After four months, the group was tested and found to have reduced their blood glucose levels to normal or near normal. Another group, given 200 micrograms per day of the mineral, showed a significant but less dramatic benefit.

These results have led even the cautious doctors affiliated with the American Diabetes Association to say, 'this is a big leap.'

But is it a leap in the right direction? I think it is. The combination of HCA and chromium, coupled with the multidisciplinary approach outlined in the next chapter, presents a powerful alternative to a drug therapy that may only forestall inevitable complications. This approach represents a holistic method of fighting this slow and silently progressive disease which does its damage with chronically elevated blood sugar.

Hydroxycitric Acid and Obesity Regulation

- Energy intake and expenditure is a finely regulated system in most animals.
- Body weight increases when energy expended goes down or intake goes up.
- Obesity causes greater risk for heart disease, diabetes, stroke and other diseases.
- Obesity treatment and diabetes prevention is helped by metabolic control.
- Hydroxycitrate (-) hydroxycitric acid, or HCA, is an inhibitor of ATP-citrase lyase.
- HCA inhibits the production of enzymes that allow fat to be formed.
- HCA curbs appetite because the brain sense that caloric intake is sufficient.
- HCA causes less eating in a dose-dependent relationship in animal studies.
- The decreased appetite in those on HCA is felt to be due to metabolic processes.
- HCA was studied extensively by Hoffmann-La Roche, Inc. by A. C. Sullivan, Ph.D. and J. Triscari, Ph.D. and L. Cheng.
- We must continue to seek to increase energy output and decrease caloric intake and decrease energy (fat) storage to adequately manage obesity and type 2 diabetes.

Adapted from *Appetite Regulation and Drugs and Endogenous Substances.* A. C. Sullivan, Ph.D., J. Triscari, Ph.D., and L. Cheng, Hoffmann-La Roche, Inc., Nutley, New Jersey, 139–160

My Approach

Patients don't like to work at wellness. I don't say this out of arrogance, but out of personal experience. After all, I too have fought my own battles with type 2 diabetes, and I know how hard it is to make changes. Nevertheless, beating type 2 diabetes calls for changes in lifestyle that aren't always easy to make. In my experience with myself and my patients, I have seen how HCA with chromium can make lifestyle changes easier.

- **HCA suppresses appetite:** Because it makes you feel full, HCA makes it easier to eat less. For many patients it eliminates even the sharpest cravings.
- **HCA and chromium increase energy without giving you the jitters:** Since both medications improve the availability of glucose, you are less likely to have the hypoglycaemic attacks that are so common in people with diabetes. These attacks leave you exhausted and frightened, sapping you of your energy.
- **HCA and chromium battle 'Syndrome X':** While HCA lowers blood fats, chromium works in tandem with it to increase insulin sensitivity. The result is a rapid change in the way you feel, because your body is getting more of the glucose that it needs, when it needs it.

These factors associated with HCA and chromium contribute to managing lifestyle changes. This is good, because the importance of diet and exercise in the war against diabetes can't be understated. I tell my patients that the fundamental thrust of treatment needs to be a multi-disciplinary approach that includes at least a new diet and daily exercise. This along with appropriate nutritional supplementation will hit the target closer than adding another medication to a long list of medications these patients are often taking every day.

I have a number of patients who have successfully adopted this approach. Below is a chart showing the improvement made by five patients who followed the Diabetes Cure. All of these patients, it should be noted, lowered their blood glucose levels to normal within four to eight weeks. This chart shows the changes in their blood cholesterol, body weight and blood pressure brought about by the Diabetes Cure after following the programme for only three months.

Initial Visit	**3 Months Later**
Case #1	
54-year-old female	
Body Weight: 24½ stone	Body Weight: 21 stone 2 lb
Blood Pressure: 128/86	Blood Pressure: 116/74
Total Cholesterol: 174	Total Cholesterol: 136
Case #2	
39-year-old female	
Body weight: 19 stone	Body Weight: 16½ stone
Blood Pressure: 116/86	Blood Pressure: 110/76
Total Cholesterol: 265	Total Cholesterol: 172
Case #3	
36-year-old female	
Body Weight: 15 stone	Body Weight: 12½ stone
Blood Pressure: 120/80	Blood Pressure: 110/60
Total Cholesterol: 260	Total Cholesterol: 199
Case #4	
59-year-old male	
Body Weight: 19 stone	Body Weight: 17½ stone
Blood Pressure: 140/78	Blood Pressure: 122/70
Total Cholesterol: 463	Total Cholesterol: 238

Case #5
62-year-old male

Body Weight: 17 stone	Body Weight: 15 stone
Blood Pressure: 148/100	Blood Pressure: 124/86
Total Cholesterol: 252	Total Cholesterol: 180

Real People

Of course, graphs don't present a personal picture of the success stories that have been achieved with the Diabetes Cure. It is important to have stories as well as statistics. Several case studies outlined in this book of people who have followed the plan and the improvements they have experienced will motivate you to try the programme for yourself.

Janice: A 'Quick Study'

Typical of patients on the diabetes cure is a 59-year-old woman I'll call Janice. She was determined to lose weight and control her high glucose level. She had been told several times over the years that she had borderline diabetes and as a result she would 'yo-yo' diet, losing and then regaining 20 pounds over and over again.

When Janice was tested at her annual physical examination, her cholesterol level was found to be 246, nearly 50 points above what it should have been. Her fasting blood sugar level was a whopping 230 mg/dl, well above the recommended 110. She was no longer considered borderline. This time she was diagnosed with type 2 diabetes.

Janice decided to educate herself on her new disease. When she found that it could be reversed, she came for counselling and the alternative treatment offered through the Diabetes Cure.

She was given an HCA formulation of 1,000-mg capsules of *Garcinia cambogia*, half of which is active substance. She was instructed to take two capsules about 30 minutes before each meal. She also kept a food diary, exercised daily and did about 20 to 30 minutes of yoga per day to reduce stress.

When she returned for her eight-week follow-up it was clear that some things had changed. She was beaming at her 11-pound weight loss, most of which she credited to the HCA and her food diary, which made her aware of all the snacking she did. Her cholesterol showed signs of significant improvement, too, dropping 36 points to 210. Most impressive of all was the reduction in her blood sugar level, now down to 111 mg/dl.

'I feel back in control again,' she said. 'I feel like I've beaten a disease that I wasn't suppose to beat.'

Gary: An Engineered Approach

Another typical diabetes cure patient is a man I'll call Gary. He is the epitome of an engineer. Having been diagnosed at a health screening fair, he confirmed the diagnosis with a fasting blood sugar test administered by his family doctor.

Knowing that most things can be fixed, he read everything he could find about the disease. With his personal computer in front of him, he designed a weekly schedule that involved eating, exercise, weight reduction and stress control.

He was prescribed HCA tablets and chromium and instructed to take them 30 minutes before each meal. As with other patients, his capsules contained 500 mg of active HCA.

After eight weeks on HCA he lost almost 12 pounds and lowered his cholesterol from 180 mg/dl to 160. His blood sugar, formerly in the high 200s, was now 97.

'I just followed the instructions,' said Gary, when he was praised for his health improvement. 'I kept telling myself that the disease is worse than the cure. That was what convinced me to stay on track.'

Bill: 'I'll Do What I'm Told'

Bill Williamson, a burly 62-year-old, was shocked to learn he had elevated blood sugar. It really should have come as no surprise. He loved sugar and had no dislike of fat, either. Every morning he started his day with about three cups of coffee, each cup sweetened with a couple of packets of sugar. He disliked the pink packaged sweetener and heard awful things about the blue one, so he stuck with the good old-fashioned 'proper' granular sugar, as he had done for years.

His visit to his doctor was just a routine check-up, until he was told that his weight of 22½ stone was literally killing him. His blood pressure was in the dangerous range and his blood sugar was 277 mg/dl. 'It must have been the donuts I ate yesterday,' he joked. But he knew that wasn't possible, since he hadn't had anything to eat for 10 hours prior to the examination. Surprisingly, his cholesterol and triglycerides were not elevated.

Bill represents a typical older patient, somewhat set in his ways, with little insight into the dramatic effects nutrition plays on health.

He had always been overweight, about 18 stone, but lately the pounds kept coming on. In fact, when he looked back at his weight, over the previous four months he had gained more than a stone.

Bill is also in the age group who usually listen to their doctor. 'I'll do what you tell me,' he said to me when I told him about HCA and what it would take to accomplish the Diabetes Cure.

He agreed that he could begin to watch the sugar, the sweets, the excess fats, and was determined to begin an exercise programme. He understood the necessity to begin slowly so as to not hurt himself or set goals that were too high. He also began taking a product with HCA in it three times daily, slightly before meals. With only a few exceptions, he remembered to take the capsules.

On his eight-week follow-up his fasting blood sugar was down to 165, while his weight had dropped and his exercise level was up to 20 minutes five times a week. Bill felt much better, and boasted that he found he could go all week without ice cream in the evening, thanks to the HCA.

But What Will My Doctor Say?

Some people are afraid of their doctors. They see them as authority figures, inflexible and all-knowing. They think they will be rejected or ridiculed by the 'white coat' if they mention a new treatment for a common disease like diabetes.

The possibility of such a response is always there, but I have noticed that even the most rigid of my peers are showing flexibility. New research has shown medical doctors that a minimal reliance on heavy-duty drugs is not only effective, but sometimes best. They have also seen the public embrace this holistic approach to medicine. Public polls have made it clear that people want to be actively involved in their own health, and this means trying a variety of alternative treatments.

Because of what doctors now know about holistic treatments, and because of the involvement that the public wants, my peers are more willing to help their patients try a different approach to healing. Indeed, the entire medical establishment is changing its feelings about holistic treatments.

Already, doctors are advocating lifestyle change instead of drugs as a way of avoiding type 2 diabetes. 'You don't have to become diabetic if you don't want to,' says Alan J Garber, MD, a guru of diabetes treatment. And I say that you don't have to stay one. The guidelines in this book will take you down a road towards healing. Don't try to do it alone. Even though you don't need a prescription to purchase HCA and chromium, you should not try to treat yourself without the medical professionals who make up your diabetes care team. Take this book with you and tell your doctor what you would like to do. Make sure you impress upon your doctor that you are ready to work hard to make your diabetes better.

I guarantee that a good doctor appreciates an interested and involved patient.

CHAPTER THREE

Nine Steps
to a Cure

Kelsey had a lot going against her when she decided to follow the Diabetes Cure. The 36-year-old secretary had a family history of diabetes, high blood pressure, heart disease and stroke. She developed type 2 diabetes in her twenties and had progressed through the oral medications and intravenous insulin until she was now using an insulin pump to control her glucose levels. She was also taking a prescription blood pressure medication that kept her blood vessels from contracting.

The combination of diabetes, high blood pressure medication and excessive body weight (Kelsey weighed 14 stone 10 lb at 5'4") left her exhausted by the middle of the day. She began to wonder what the years ahead would be like, and saw the answer to her question in her own family. Her father was nearly housebound by his own diabetes and high blood pressure, while her mother was depressed about being in the early stages of type 2 diabetes and getting worse. One of her uncles had suffered for years from diabetes and had recently had a stroke which had left him partially disabled.

Not wanting her own condition to worsen, Kelsey began to follow the nine steps of the Diabetes Cure. 'I want to become engaged again in my health care,' she declared, when I asked her why she had decided to pursue an alternative treatment for her diabetes.

Kelsey began a programme of HCA and chromium supple-mentation. She also began to watch her diet very carefully, avoiding fried and fatty foods with a new-found diligence. She began to earmark times on her day planner for when she was going to walk, deciding to exercise for 20 minutes before her shower in the morning and again at midday.

She followed the other parts of the nine-point programme, too, including taking several vitamins as well as finding medi-tation time for herself to break the stress spiral.

Within eight weeks she had lost 12 pounds. More important, her cholesterol had dipped below 210 and her glucose level to 98. She has continued to stay on the programme and is using much less insulin from her pump. She has now begun to talk to her endocrinologist about how she can get rid of her pump and return to, as she puts it, 'an untethered life'. Her health status has improved, as have those of many others, by following the nine steps of the Diabetes Cure:

1   See your doctor for a thorough consultation and examination.
2   Follow your doctor's guidance and take your HCA and chromium.
3   Prepare yourself.
4   Exercise daily, passively and actively.
5   Eat a healthy diet and don't stray.
6   Fight antioxidants with vitamin supplements.
7   Stay stress-free.
8   Monitor your progress daily.
9   Pursue a positive attitude.

Although this may seem like a lot to do, this advice must be followed to foil a disease like diabetes, even if you decide to use prescription medication.

Unlike the 'old days', when people took medication and did little else to fight chronic disease, more people are willing to be actively involved in treating their condition and preventing complications. A study of baby-boomers diagnosed with diabetes found that they were 50 per cent more likely than older patients to discuss specific ways to prevent complications. That means baby-boomers are more proactive about their health. Although they become angry, frustrated and depressed at hearing about their disease, the study says they are more likely to do something about it.

HCA and chromium are only the beginning of the Diabetes Cure. They are an important part of a nine-step programme that attacks this insidious disease from all sides.

Undertaking any 'project', even a health project, is fun and exciting at first. When first hit with the diagnosis of 'diabetes', or 'borderline diabetes' or just getting into the 'health of it', many of my patients are very keen. Reading, charting, tracking and monitoring are actually reported to be 'fun'. The honeymoon wears off for most of us, though, as the dreary days of winter set in, or the glucose meter runs out of strips, or we look at a cake and think, 'just one piece won't be so bad.'

Diabetes, whether severe, mild or borderline, must be carefully and intently dealt with. This is not a forgiving medical problem. This disease takes what it wants when no one is looking. Before beginning the nine-step programme, sit quietly for a few minutes and perform a reality check. We know a lot about the nature of this disease and how various components interact with it. This is a disease requiring great respect and perseverance, but it is a controllable disease.

The nine steps outlined in this chapter will guide you to taming diabetes. The nine steps are designed to be realistic and attainable and will work for you if you expend the effort to follow them.

Read through the details of these steps carefully and answer truthfully, 'Can I follow this step?' If you can, you will be following a path to self-improvement and healing.

Step 1: See Your Doctor for a Thorough Consultation and Examination

Current medical guidelines urge all adults age 45 and over be tested for diabetes. If the test results are normal, they should be tested again at three-year intervals. Anyone with the risk factors listed below – even children – should be tested more frequently:

- those who are 20 per cent above ideal body weight
- women who have given birth to babies weighing more than nine pounds
- anyone with parents who have had diabetes
- people with 'high risk' ethnic backgrounds (Afro-Caribbean, Asian, Hispanic, Native American)
- people with hypertension (blood pressure readings of 140/90 or higher)
- people with elevated blood lipids
- people who have previously been found to have impaired fasting glucose or impaired glucose tolerance.

If you have diabetes already, you are probably very familiar with the following tests and how to interpret their results. These tests give you a baseline from which to work, because they reveal your blood glucose levels.

FIGURE 3.1

Progression of Diabetes

Factors Involved in Acquiring Diabetes

Genetic predisposition, environment, nutrition, obesity, lifestyle.

Causes of Impaired Glucose Tolerance

Increase in insulin resistance

Increase in insulin levels

Transient increases in glucose levels

Increase in total cholesterol

Decrease in HDL cholesterol

Increase in triglycerides

Causes for the Onset of Diabetes

Atherosclerosis – cardiovascular disease

Elevated blood glucose levels consistently

Elevated blood pressure

Further Progression of Diabetes

Retinopathy – Eye disease

Nephropathy – Kidney disease

Neuropathy – Nerve damage

Late-Stage Complications

Blindness

Renal (kidney) failure

Coronary artery disease

Amputation

Other organ damage

Random Plasma Glucose Test

This is the simplest way of detecting diabetes. It is usually the first test done by a doctor who suspects diabetes as the cause of symptoms like frequent urination, sudden weight loss, blurred vision and extreme thirst. It is simply a blood test that measures the amount of glucose in your blood.

To conduct the test, a doctor will draw blood and have it tested in a glucose meter. The test relies on the fact that diabetes keeps your blood sugar levels above normal most of the time.

Results
Test results of 200 mg/dl or higher mean that it is likely you have diabetes.

Although this test is very effective in detecting diabetes, I prefer the following one.

Fasting Plasma Glucose Test

Of the several possible methods of diagnosing type 2 diabetes, I recommend a fasting plasma glucose test. This method of screening can be done by your doctor. A very similar version of this test can be performed at home on a glucose meter, which is available at most chemists. Whether done at home or in the doctor's surgery, this test requires that you not eat food or drink anything but water for a full eight hours before taking it.

If a doctor is involved, he or she will draw blood to be tested by an in-house glucose analyser or an independent laboratory. Your doctor will likely order other tests at the same time to get a complete profile of your blood.

If you do this test at home, you will have to calibrate the glucose meter to ensure the accuracy of the test. It is essential that you follow the directions provided with your glucose meter, because the tests are very specific. You will also have to draw your own blood with a needle stick in the finger.

Results

Test results of 110 mg/dl or under is considered normal. Levels over 126 mg/dl indicate type 2 diabetes.

The criteria for diagnosing type 2 diabetes indicates that the test must be confirmed by a second reading on a different day.

Two-hour Glucose Tolerance Test

The two-hour test is recommended by the World Health Organization and is performed by drinking 75 grams of glucose dissolved in water. Two hours later, blood is drawn and the plasma glucose concentration is measured.

The two-hour test replaced the previously used eight-hour test, which called for an eight-hour fast followed by the glucose cocktail. Blood was then drawn six times over the next six hours. A one-hour version of this test is recommended by some experts for pregnant women who suspect they have gestational diabetes.

Results

A non-fasting patient with a value of 200 mg/dl or greater on two separate occasions of taking the two-hour glucose tolerance test is diagnosed as having diabetes.

Any of the above tests will give you a baseline from which to improve your blood glucose levels.

Step 2: Follow Your Doctor's Guidance and Take Your HCA and Chromium

If you have type 2 diabetes or borderline diabetes and want to follow the Diabetes Cure, it is important to talk to your doctor about how to make the gradual switch from the pharmaceutical methods of blood glucose control you may currently be on to this more natural method.

If you have borderline diabetes or have had diabetes for only a short time, it will certainly be easier to be weaned off diabetes drugs than if you have had the disease for a long period of time. Be sure to get an evaluation from a medical doctor, a doctor of osteopathy or an endocrinologist. They will want to administer a complete physical examination and provide some biofeedback in the form of a blood test.

The test I use to monitor my patients is a fructosamine assay. This test is able to 'average' your glucose levels over the previous three-week period to see how consistent your blood sugar levels have been.

Another excellent test for monitoring your progress is a glycohaemoglobin test, or haemoglobin A1-C, which monitors glucose control over the previous three-month period.

I prefer the fructosamine test because it allows me to look at a two-week 'snapshot' of blood sugar levels and make medication or other changes faster. It also means a shorter period before patients can have some feedback, which most patients like since they want to know how well they are progressing.

I cannot emphasize enough: Do not stop or reduce your levels of medication without medical supervision.

FIGURE 3.2

Diagnosing Diabetes

Diagnosis	Fasting Glucose	Random Glucose
Normal	Less than 110 mg/dl	Less than 140 mg/dl
Impaired glucose tolerance (GST)	Greater than or equal to 110 mg/dl and less than 126 mg/dl	Greater than or equal to 140 mg/dl and less than 200 mg/dl likely
Provisional diabetes (must be confirmed)	Greater than or equal to 126 mg/dl	Greater than 200 mg/dl

Criteria for diagnosis of diabetes mellitus:

- Symptoms of diabetes plus casual plasma glucose concentration greater than or equal to 200 mg/dl. Casual is defined as any time of day without regard to time since last meal.
  OR
- Fasting plasma glucose (FPG) greater than or equal to 126 mg/dl. Fasting = no food for eight hours.
  OR
- 2 hours plasma glucose greater than or equal to 200 mg/dl during an oral glucose tolerance test. Glucose tolerance challenge is using a 75-gm anhydrous glucose in water.

A thorough history and examination should be performed by your health care provider upon your request. During the initial visit, you will be asked to provide some background medical information which your doctor will review to determine if there are any risk factors for diabetes and other diseases.

This history can uncover symptoms that alert the doctor to other problems needing to be addressed for your total care. If you have diabetes, the history will help determine if there are any other problems which need attention.

There are some symptoms and information with which you should provide your doctor, whether or not you are asked about them. Being proactive and providing this information will help your doctor evaluate your risk:

- symptoms of diabetes, such as excessive urination, thirst, periods of light-headedness, or hypoglycaemia
- prior screening test results for diabetes, such as fasting or random glucose tests
- prior special tests for diabetes, like oral glucose tolerance tests
- prior glycohaemoglobin results or fructosamine tests (glycoprotein tests)

- food intake patterns, weight history, gains or losses, diets, current nutritional status
- prior diabetes treatment programmes or educational classes
- current treatment programme for diabetes or weight management
- physical fitness level, exercise history, current exercise regime
- complications of diabetes if any, when, where and how severe
- current and previous infections, especially of the skin, feet, teeth, urinary tract and chest
- recurrent yeast infections or thrush (oral yeast infections) in the mouth and throat
- complications of diabetes of the eyes, nerves, heart, skin and feet, intestines, kidneys and bladder, blood vessels in the arms, legs and brain; sexual problems
- current and previous medications used to treat illnesses, both short and long term
- risk factors for heart disease like smoking, high blood pressure, obesity, high cholesterol and high triglycerides, and family history of heart disease
- family history of diabetes and other 'hormone' diseases
- if female – history of large deliveries (babies weighing 9 lb or more), toxaemia, stillbirth, excessive amniotic fluid
- any other factors which may cause you to be at increased risk for diabetes or would enter into the treatment approach to diabetes like lifestyle, culture, economic status, education, psychological issues, level of support at home, time required for self-management of this disease.

This is a long list, but one that will provide your doctor with the information he or she needs to give you the best possible care.

This assessment should be conducted yearly. If new problems arise, your doctor should be informed. Of course, if the problems are of an urgent nature, notify your doctor immediately.

Once you and your doctor review the history, he or she will perform a physical examination. A 'complete physical' means

different things to different people, even in the medical profession. To evaluate you for diabetes, a physical examination
should include the following:

- height and weight and calculated Body Mass Index or BMI. The
  BMI is a useful tool to predict health risk. People with a BMI at
  25 or greater have an increased statistical risk of medical
  problems associated with weight. A BMI of 25 to 27 is termed
  'overweight', while the higher BMI values indicate a diagnosis
  of 'obesity'. Your health care provider should be able to help
  you determine your individual BMI.
- a body composition is recommended but is not available in
  most surgeries. This test measures the amount of body *fat* (fat
  tissue) as compared to body *lean* (muscles and bones). The
  result is usually a percentage of fat.
- sexual maturation (for children and adolescents)
- blood pressure, pulse rate, respiratory rate and temperature
- eye exam, both screening reading and with an
  ophthamoscope. A formal eye exam by an ophthalmologist or
  eye specialist is preferred so the eyes can be dilated and a
  detailed exam performed.
- examination of the ears, nose and throat including the
  thyroid gland
- examination of the lungs, heart and abdomen
- examination of the pulses in the arms, legs and neck
- examination of the nervous system by touch and reflexes
- examination of the skin over the body, especially the hands
  and feet
- other areas may be examined based on age and gender like the
  rectal, pap smear and breast exam for women, rectal, prostate
  and testicular exam for men. These additional tests should be
  performed according to screening guidelines but can be
  requested by you prior to the exam.
- various cancer screening tests may also be recommended
  based on your age and gender, such as a mammogram for

women and a sigmoidoscopy or haemmoccult testing to look for colon cancer for both sexes.

The clinical examination and the history are added to a comprehensive evaluation of blood tests which are performed by a laboratory. Basic blood and urine testing will look at over 50 values of different things occurring in your body, including:

- fasting plasma glucose levels
- total cholesterol
- HDL (good) cholesterol levels
- LDL (bad) cholesterol levels
- triglycerides
- blood chemistry including sodium, potassium, creatinine, blood urea, nitrogen and others
- urinalysis looking for elevated glucose, protein and ketones and infections
- serum albumin (protein)
- thyroid tests and thyroid-stimulating tests.

Other Tests

Other tests might include ECG or electrocardiogram to look at the heart, chest X-rays if you smoke. If, based on your medical history, the doctor suspects anything else should be looked at, other tests could be indicated as well.

The examination usually concludes with a summary from the doctor on the findings of the exam. Sometimes doctors, as I do, will ask you to come back in a few days to go over the results in detail. Since there is so much data to bring together, and the results of the tests take some time, it is usually more beneficial to return for a summary visit.

After careful review of the results of your history, examination and testing, your doctor will work with you to develop a plan for improving your health, or keeping you on the right track if you are without problems. With your baseline health established, you will now be ready to take full advantage of Step 2, which is taking the HCA and chromium supplements.

Correct Dose, Right Time

The two supplements, HCA and chromium, combine to curb your appetite, metabolize fat and, most important, increase your sensitivity to the insulin produced by your own pancreas. To achieve the maximum benefit of these supplements you need to take the correct dosage at the proper time.

How Much Should I Take?

I recommend taking about 1,500 to 5,000 mg of *active* HCA per day and about 1,000 mcg of chromium per day. These should be taken three times per day, one hour before each meal, with two-thirds of a pint (16 fluid ounces) of water.

I emphasize *active* because only about half of the garcinia contained in the supplements is actually HCA. The remainder, as with any herbal preparation, is plant matter that is not used by the body. This difference between garcinia content and active HCA has been taken into consideration in the chart below.

Research has shown that it is virtually impossible to take too much, or 'overdose' on HCA. Researchers liken its toxic effects to citric acid, the substance that gives oranges, limes and other citrus fruits their tangy taste. If you were to take very large quantities of HCA you would certainly suffer from stomach aches or other gastric disorders, but that would likely be the extent of your discomfort.

I generally recommend that patients with delicate digestive systems take the lowest recommended dosage for several days before increasing the dosage, to give their stomachs time to adjust to the new supplements.

Within a few days your glucose tests should reflect the effects of the HCA/chromium supplementation and the other steps in the Diabetes Cure. Most of my patients see their glucose levels rapidly approaching normal. Those who are taking medication are able to take less and less, which I recommend doing only after they talk to their doctor.

Even patients who are not able to stop taking medication entirely are able to reduce the amount they have to take. Almost all patients who follow the steps in the Diabetes Cure report more energy and vitality and, for the ones who need it, weight loss.

Step 3: Prepare Yourself

Making lifestyle changes is not easy, and the full effect of the Diabetes Cure cannot be attained without making changes in the way you live. As you read the rest of this book, ask yourself these questions:

- Am I willing to exercise at least 30 minutes every day?
- Am I able to put stressors in second place behind my good health? Will I apply the lessons in Chapter 7 on stress control to: events happening at work? Events happening at home? Emotions in my personal relationship? Bad feelings I might have about myself? Read Chapter 7 and see if you are willing to avoid stress in your life. Surprisingly, many people aren't or can't for one reason or another.
- Am I willing to watch my diet? Will I eat only 'safe' foods and not cheat?
- Will I take vitamins and minerals on a daily basis?
- Will I take the prescribed amount of HCA/chromium?
- Will I monitor my blood and see my doctor regularly for feedback?

If the answer to any of these questions is 'no,' then your Diabetes Cure programme will be compromised. Remember, medication alone won't stop the progress of diabetes.

Step 4: Exercise Daily, Actively and Passively

Exercise is the easiest way of making your body receptive to insulin. Why this is true isn't understood, but every diabetes treatment programme calls for exercise as one of its major components.

I break exercise down into 'active' and 'passive' categories. Active exercise involves continuous and concerted movement like walking, running or bicycling. It used to be thought that exercise had to leave you breathless and sweating before it did any good. A number of studies have shown that any exercise, no matter how gentle, can be of benefit. Sometimes it is recommended that you take at least 20 minutes of exercise three times per week. I recommend 30 to 60 minutes of active exercise every day. The reason? Every day is a new day when it comes to glucose control, especially if you want to keep your levels in a narrow range. Consistent and daily exercise will help to moderate your glucose levels.

Passive exercise is the calories burned from random acts like climbing the stairs instead of taking the lift, or parking your car at the far end of the car park. This is an easy way to burn extra calories and improve your glucose control.

Step 5: Eat a Healthy Diet and Don't Stray

This is the most difficult step for most people, largely because eating is one of our most accessible and acceptable forms of pleasure. Not surprisingly, the need to change their diet is one of the greatest concerns that my patients have when I tell them that they have diabetes. 'Does this mean I can't eat ice cream anymore?' they ask fearfully. Or, 'Will I have to quit using sugar in my coffee?'

The truth is that most people with diabetes *will* have to change what they eat, because they probably are not eating a healthy diet to begin with. Most of my patients consume far too many fats and sugar and have gained significant amounts of weight as a result. The added weight coupled with too little activity is probably what got them to the point of developing type 2 diabetes.

The good news is that healthy eating for a person with diabetes is no different from healthy eating for a person without it. The Diabetes Cure calls for eating from a wide variety of foods, including carbohydrates, proteins and fat. It doesn't mean you have to give up your ice cream or your sugar in your coffee, it might only mean that you will have to eat less of these foods.

There is great benefit in following the advice in this book's diet chapter. Just as there are complications to diabetes, there are complications to poor diet. Diabetes, of course, is one of those complications, but others include stroke, high blood pressure, joint problems and lack of energy. By eating a healthy diet you will not only help heal your diabetes, but will improve your overall health as well.

The healthy eating advice in this book is aimed at:

- helping you reach and maintain a healthy weight
- keeping your blood glucose within its target range
- finding a meal plan that you like and are willing to live with
- letting you know how much flexibility exists in a healthy diabetes meal plan.

Following a diet plan is important for the person with type 2 diabetes. Making you miserable is not my goal, especially since you are unlikely to stay on a regular diet plan you don't like. The dietary advice in the Diabetes Cure will help you select the foods you want to eat.

You will be amazed at what eating reasonably can do for your glucose control. In fact, the act of combining Steps 4 and 5 can be powerful medicine in and of themselves. I have a number of patients who have been able to control their high glucose using diet and exercise alone. Although it is impossible to say exactly what causes insulin sensitivity, it is clear that 'Syndrome X' is closely linked to body fat and inactivity.

The graph below shows the blood profile of three patients who managed their glucose for three months using only diet and exercise. In addition to lower glucose levels, notice the lower cholesterol and triglycerides, too.

Fasting Plasma Glucose Tests

	Total Cholesterol (Under 220)	**Triglycerides**	**Glucose** (Under 110 mg/dl)
Patient #1			
First month	529	3058	206
Third month	286	539	133
Patient #2			
First month	219	543	126
Third month	163	96	74
Patient #3			
First month	207	211	148
Third month	182	131	98

This graph clearly illustrates that type 2 diabetes is a form of 'hoof and mouth' disease – too much food in the mouth and not enough exercise on the hooves. That is why Steps 4 and 5 are so important to the Diabetes Cure.

Step 6: Fight Antioxidants with Vitamin Supplements

The only people who used to be criticized more than those who took vitamin supplements were doctors who recommended taking vitamin supplements. Recommending them for diabetes or any of its complications was almost heresy.

Now, scientists can't seem to say enough about the benefits of antioxidants. For people with diabetes, they may be indispensable. Antioxidant-containing substances such as vitamins A, C and E and alphalipoic acid may slow the development of eye, kidney and nerve complications as well as atherosclerosis, as reported at an American Diabetes Association symposium.

Of these, vitamin E, commonly found in vegetable oils as well as wheat germ, nuts and green leafy vegetables, is the most promising. Vitamin E appears to reduce low-density lipoprotein (LDL), so-called 'bad' cholesterol. In a trial at the University of Texas, researchers gave healthy patients 120 times the recommended daily average of vitamin E for eight weeks. It significantly reduced LDL oxidation and other processes that contribute to atherosclerosis.

A pilot study at the University at Bugingen in Germany found that alphalipoic acid, another antioxidant, cleared glucose from the blood more quickly while improving insulin sensitivity and glucose tolerance.

A Stanford University study in the June 1996 issue of the *American Journal of Clinical Nutrition* found encouraging evidence that vitamin A supplements of three to five times greater than the recommended daily average may enhance the body's ability to move glucose from the blood into the cells. A fat-soluble vitamin, A is found commonly in orange fruits and vegetables, green leafy vegetables and liver, fish liver oils, whole and fortified milk and eggs.

The findings of recent years have changed the negative attitude medicine has towards vitamin and mineral supplements. Although we still don't know all the subtleties of supplements, it is obvious that they should be part of a healthy anti-diabetes programme.

Step 7: Stay Stress-free

Although it is doubtful that stress can cause diabetes, there is little question that it can make it worse. It does this in two ways, and one of those is mental. When we are exposed to stressful situations, many of us have a tendency to eat or drink too much food or alcohol. Many experts think we consume extra calories because they bring us comfort in these times of distress. Some people try to overcome their stressful situation by ignoring the world around them and becoming 'couch potatoes'. And some of us do both, sitting in front of the television with a beer in one hand and chips in the other, raising our glucose levels to new heights.

Another way diabetes can be worsened by stress is by the changes in physiology it brings. Stressful situations like traffic jams cause your body to release adrenalin into your bloodstream, triggering a 'fight or flight' response. This causes your body to produce more glucose that your insulin-resistant body tissue may be able to process. The results are elevated glucose levels and a tense, even frightened feeling that can last a long time.

There are many ways of avoiding stress, especially if you know the techniques to stop it before it sneaks up on you. Chapter 7, 'De-Stressing for Diabetes Care', covers the best, most effective methods of staying stress-free.

Step 8: Monitor Your Progress Daily

Monitoring your progress on a daily basis is an important part of diabetes care. The goal of daily monitoring is to keep your blood glucose under tight control and as close to normal as possible.

The importance of tight glucose control has been emphasized for many years by doctors who treat diabetes. Recently a major study showed just how important this is.

Called the Diabetes Control and Complications Trial (DCCT), the study followed 1,441 people with diabetes. Half of them checked their glucose levels two or three times per day. The other half tested their glucose levels four to seven times per day.

The results surprised even some of the doctors who have long been advocates of tight glucose control. They found that intensive control of glucose reduced the risk of complications resulting in eye disease by 76 per cent, nerve disease by 60 per cent, kidney disease by 50 per cent and cardiovascular disease by 35 per cent.

Keeping tight blood glucose control means frequently monitoring your blood with a glucose meter. These devices are about the size of a pack of cards, and give read-outs of your blood sugar levels in less than a minute. Based on the results, you can quickly adjust diet, exercise or medication to keep your blood glucose levels within an acceptable range.

Using a glucose meter involves pricking your finger with a lancing device to draw a drop of blood. The blood is then put on a paper test strip and inserted into the meter.

Consumer Reports magazine tested a number of glucose meters for convenience, accuracy and consistency. Among the top 10 were the following available in the UK:

- Lifescan One Touch Profile
- Lifescan One Touch Basic

Among other products listed in this report were those manufactured by Bayer and MediSense.

Please note that glucose meters are not available on the NHS. They range in price from £25 to £200, with most around the £35 mark. Please consult with your doctor before buying one.

Tight Control

For the purpose of the Diabetes Cure, tight blood glucose control means keeping your blood glucose level in this narrow range:

Fasting Levels

Normal	less than 110 mg/dl
Acceptable	80—120 mg/dl
Needs improvement	Less than 80 and greater than 140 mg/dl

Bedtime Levels

(new in 1998 to prevent hypoglycaemia during the night)

Normal	less than 120 mg/dl
Acceptable	100—140 mg/dl
Needs improvement	less than 100 and greater than 160

Postprandial Plasma Glucose

(taken 1.5—2 hours after eating to see how food affects glucose levels)

Normal	less than 140 mg/dl
Acceptable	less than 180 mg/dl
Needs improvement	180—235 mg/dl

Glycohaemoglobin

(test that measures the amount of glucose that sticks to red blood cells. Also known as HbA1c or GHb)

Normal	less than 4.0—6.0%
Acceptable	6.0%—7.0%
Needs Improvement	greater than 8.0%

Although daily self-monitoring of blood glucose is important for people with type 2 diabetes, especially those on medication, the number of times you test yourself varies. Research published in *Clinical Diabetes* recommends testing enough times to reach your glucose goals. For most people, three times per week will be adequate, for others four to seven times per day might be called for. My recommendation is to test yourself daily in the first several weeks of the Diabetes Cure and then taper back when you have established glucose control.

I also recommend glycohaemoglobin testing at three-month intervals, or better yet fructosamine testing at three- to four-week intervals. These tests assess the results of your glucose self-testing regimen and indicate whether a treatment plan is working.

FIGURE 3.3

Blood Glucose Monitoring

ALL diabetes patients should monitor their blood glucose levels.

If you are controlling your diabetes with diet or oral medications:

Check your blood glucose levels before breakfast and before dinner, at least two or three days each week.

If you are using insulin to control your diabetes:

Check your blood glucose 4 times per day – before all meals and bedtime, at least two to three days each week or more.

Less testing is needed when test values are in a consistently normal range.

More frequent testing is needed when levels are fluctuating or elevated.

Your physician may request that you check your levels more or less frequently.

Always calibrate your glucose monitor by following the directions supplied with the unit.

Too High, Too Low

If you test yourself and find that your glucose levels are too high, you may want to lower them by increasing the amount of exercise you are taking or by eliminating some carbohydrates from your diet. See Chapter 5 for a detailed explanation of diet's role in glucose control.

It is also possible and very common for stress to elevate your glucose levels. Stress can come from a number of factors, including:

- illness
- infection
- being upset
- being angry
- being frightened.

If you think that any of these factors is leading to elevated glucose levels, see Chapter 7.

If your glucose level is too low, you should evaluate your diet:

Are you eating too many simple carbohydrates? Are you eating too few calories in general? Also assess your exercise habits, since too much exercise can cause glucose levels to drop. Also examine your medication intake, since taking too much can cause blood sugar to be removed from your blood at a very rapid rate.

Although tight glucose control is by far the best way to avoid the complications of diabetes, you will probably have more episodes of hypoglycaemia than you would without tight control. When you feel the symptoms of hypoglycaemia (moodiness, shakiness, dizziness, numbness in the arms and hands), test your blood glucose levels as soon as you can, or eat a small amount of carbohydrate to alleviate the feelings.

The American Diabetes Association (ADA) recommends basing carbohydrate intake on your blood glucose levels. On the average, 5 grams of carbohydrates raises blood glucose about 15 mg/dl. Your blood glucose goal is about 120 mg/dl. Below is the chart the ADA provides for raising your blood glucose levels. Note: 1 teaspoon of sugar equals about 10 grams of carbohydrates.

If your blood glucose is:	Eat this much carbohydrate:
under 40 mg/dl	30 grams
40 to 50 mg/dl	25 grams
51 to 60 mg/dl	20 grams
61 to 80 mg/dl	15 grams
Over 80 but have symptoms	5 to 10 grams

Step 9: Pursue a Positive Attitude

One of the greatest tonics of all is self-encouragement. To accomplish this, stay aware of your goals. Of course, this can be difficult.

There are so many pressures in life. Add to those coping with a chronic disease like type 2 diabetes, and you have an additional set of pressures most people don't have.

When pressures mount up, most of us fall back into familiar patterns of behaviour because they are, well, comfortable. Many of my patients with diabetes suffer setbacks when they revert to eating (usually overeating) as they used to, or failing to exercise. Some forget to take their medication because they can't get accustomed to a new routine, or they become overly stressed about events in their lives instead of remembering that it may have been poor coping techniques that got them into the lifestyle habits leading to their diabetes to begin with.

Since diabetes is a silent disease, especially in its early stages, people often don't necessarily feel better as a result of their lifestyle changes, because they didn't really feel bad to begin with. When this happens it is difficult to make changes. After all, they ask, if I don't feel really awful, why should I change my lifestyle anyway? Thinking like this leads many people to wait until it is much later in the disease process before they deal with their diabetes. The longer you wait before involving the healing process, the longer it takes to heal.

It is difficult to break old routines and replace them with healthy, new ones. New and healthy routines form the back-bone of the Diabetes Cure. There isn't a pill made that will slow, stop or reverse diabetes alone. All medications take the concerted effort of the eight steps above.

So how do you stay motivated throughout this treatment?

One method I recommend to my patients is to keep a daily record that includes their blood glucose count, what they eat, how much they exercise and how they feel.

I have included a daily progress chart in the Appendix to this book. Easy to copy and fill out, the progress chart is important because it adds that much-needed daily motivation.

The support of family and friends will also help you stay motivated. Make sure they understand what you are trying to do by following the Diabetes Cure.

- Have family and friends read this book, so they will understand its multi-disciplinary approach to fighting this disease.
- Ask them to be supportive by encouraging you towards your goals instead of away from them. This is especially true in the dietary arena, where they might try to encourage you to have one more serving of dessert, or an extra helping of mashed potatoes.
- Ask them to participate in various parts of the programme. Although they might not have diabetes, maybe they could benefit from a change in eating habits or an exercise programme. Remind them that it is easier to make lifestyle changes if more than one person is involved.

People with diabetes have more ups and downs than the general population. Much of the depression that comes from diabetes is the realization that you have to cope with a chronic disease. It is a known medical fact that people with chronic diseases have more depression than those without.

Depression is more common among people with diabetes, largely because of the physiology of the disease. I call this 'blood glucose depression' largely because it can be brought on by variations in your glucose levels.

It is important to be aware that many times the 'blues' you are experiencing can be eliminated by bringing your glucose under control. Fighting back against this depression by following these steps will help you win against these bouts of the blues and make you feel in charge of your disease.

Does the Diabetes Cure Work?

The Diabetes Cure is a unique approach to the treatment of type 2 diabetes. It offers a non-prescription medication and steps for putting you back in control of your life. At the very least, the vast majority of patients with type 2 diabetes who follow this programme will show significant improvement in their health. Of course, this programme works best with borderline diabetes (people on the verge of getting the disease) and those who have not had the disease longer than five years or have not progressed to taking insulin. It does however have a positive effect on virtually *everyone* who follows it.

Here is how the programme has worked for some people:

J R: 'So Many Things to Do'

One of our patients, J R, is thrilled with the results of the Diabetes Cure regimen. He is a 52-year-old Afro-Caribbean with several of the common diabetes complications. Early in 1997, he decided to take a more active role in his diabetes management after a visit to our clinic.

Being a busy man, he freely admitted that focusing on his health was usually not high on his priority list. 'So many things to do all the time and not much time to do the things that I know I should,' he said.

He agreed to become more involved with his disease and was very open to the learning process.

In March 1997, a screening blood sugar revealed a glucose level of 299, well over the desired range of 65—110. His liver enzymes were also elevated and his total cholesterol showed a high level of 259. His 'good' cholesterol, or HDL cholesterol, which should be at a minimum of 45 or well over, was only 42. His 'bad' cholesterol was 180, higher than the recommended level of 130.

I ordered a test that would show us how well his overall diabetes control had been in the past three months. The measurement of this test provides a glimpse of how long the glucose level has been high, by measuring its effects on glucose binding to red blood cell components, namely haemoglobin. This is called a glyco (for glucose) haemoglobin (for the protein in the red blood cells which carries oxygen) test. The glycohaemoglobin test, sometimes referred to as a haemoglobin A1-C (HbA1c), is useful in following how well our interventions against diabetes is working.

His test came back an 8.8. The normal values for the test is 4.2—6.5.

With a knowledge of how much he would have to progress, J R began taking the Diabetes Cure. He increased his physical activity and visited with our clinical nutritionist, Soheyla Radfar, PhD, who found that J R was eating an excessive amount of deep fried foods, refined sugar and red meat. He was asked to try to improve his choices each day.

Dr Radfar taught him about the foods he was already eating and explained which ones were good for him and which ones weren't. Suggestions were made to alter the way the food was prepared to keep the flavour and get rid of the fat.

Things began to change in J R's life. He began to lose weight, to drink much more water than before, and to eat smaller meals four to five times a day. He didn't feel hungry.

He was walking 10 to 15 minutes, four to five times a week and was changing his daily routine to fit in more exercise. His wife even joined him on several of the walks when she could. The walks helped him 'unwind' from his hectic day, and pay attention to his body. Now he trimmed the fat off his meat and ate a piece of fruit on the way home from work instead of stopping for a milkshake. He was changing his lifestyle, and it wasn't as hard as he thought it would be.

When he came in for his check-up, J R was pleased to see that his blood pressure had dropped. The results of his blood test showed a man who had changed for the better. His glucose level was now 105, well within normal, his glyco-haemoglobin test, which was previously 8.8, was now 7.1, not normal but quite improved. His total cholesterol level was down to 192, while his LDL or bad cholesterol was down to 135 and his risk factor for heart disease was calculated to be 4.56, well below the 6.12 it had been previously. J R was pleased indeed.

J R continued to follow the suggestions for the Diabetes Cure and his glycohaemoglobin level was found to be 5.4 only four months after he began to focus on his health.

Sean: 'I Feel Better'

Sean is a 57-year-old man with non-Hodgkin's lymphoma which has been complicated by hypertension and diabetes for years. His blood glucose levels were always elevated, sometimes very high. He had a difficult time emotionally with the daily fluctuations and felt as if nothing he did could control the varied results. He was discouraged with these fluctuating numbers and at times felt, 'What's the use, no matter what I do, I keep getting bad results.'

Diabetes can be very discouraging for many people who react like Sean. They try to become involved in their disease but the feedback they get keeps telling them that, no matter what they do, the glucose numbers will still be off.

Sean began searching for an alternative approach to help him with his diabetes when he found our clinic. He appreciated the traditional medical approach to his cancer, but wanted to help this traditional approach by including alternative treatments for his diabetes. He heard a lecture I gave at a local natural health food store on the values of HCA for those interested in controlling their diabetes in a natural fashion, and was intrigued by what he heard.

He was seen by our clinical nutritionist, who explained the pros and cons of HCA. After that, he was determined to begin using the fruit extract. He was given six tablets of a form of HCA that contained 500 milligrams of garcinia yielding 250 mg of active HCA, and told to take two tablets three times a day. The formulation also contained 75 mcg of chromium. In addition to the HCA and chromium, the supplement contained vitamins B_2 (riboflavin), B_3 (niacinamide), B_5 (pantothenic acid), manganese and L-carnitine.

By his report, within just the first two weeks he began to 'feel better'. When questioned in detail about what this meant to him, he stated his energy level was improving, his eyesight was improving and his glucose had fallen an average of 10 points on routine testing.

He increased his exercise, which he was now able to do with his renewed energy levels. He has continued to improve by using the HCA/chromium compound and compliance with the other steps in the Diabetes Cure.

Sean's case points to the fact that diabetes does not frequently stand alone as a single disorder. Type 2 diabetes frequently comes along with a variety of disease states and can actually be missed altogether to an untrained eye or to the health care provider who is focused on the presenting problem only. High sugar levels, as seen with type 2 diabetes, can slow healing and interfere with recovery from a variety of conditions. A problem with glucose handling should always be in the back of the minds of patients and their doctors.

Mike: 'Full-blown' Trouble

A patient named Mike illustrates a classic case of full-blown type 2 diabetes out of control. Arriving at our surgery with a blood sugar of 351 mg/dl, Mike decided to face his problem head on.

'I never really thought about it much,' he said. 'I just tried to watch my sugars sometimes. If I didn't feel bad, I thought I was OK.'

This is such a typical, yet unfortunate, error in a patient's perception of his disease. Diabetes doesn't knock on the door and announce itself with symptoms as so many other diseases do. Diabetes waxes and wanes, causing trouble undetected. Mike realized that testing was the only way he could tell if his values were high.

My nutritionist instructed him in the essentials of a good diet for people with diabetes and what food choices would push his blood sugars levels up. She described the actions of HCA and chromium, and the benefits of these compounds.

Mike declared that he was determined to focus more on his health and devote at least 30 minutes a day to active exercise, something he had never done.

His focused energy paid off. Four weeks later, we all rejoiced at his progress. His glucose fell to 190 mg/dl, down from his initial 351 mg/dl, his exercise was consistent and his food choices were dramatically improved. Best of all, he felt great!

Mike has 'learned to live better' through the Diabetes Cure and has continued to improve by following the steps of the programme.

These are just three patients who have been helped by the Diabetes Cure. Results are seen after anywhere from four weeks to three months, depending upon the severity of the disease and how well the patient complies with all of the steps. Although HCA and chromium have both been shown to work independently of the programme, they are far more effective when used in conjunction with all the other steps.

Darlene: 'I Just Want to Avoid It'

Darlene is a 60-year-old patient who wants no part of 'getting old'.

'I don't have time for it,' she says with a devilish grin. 'I want to do everything as naturally as possible, not that I have anything against the prescription stuff, I just want to avoid it if I can.'

Darlene said she was an only child who usually didn't take 'no' well. When she was diagnosed with type 2 diabetes and told she had 'no' choice but to go on medication, she decided to come to me to see what kind of natural regimen she could follow.

Our nutrition staff went to work, outlining her daily recommended exercise requirements, the foods she should eat more of and the foods she should avoid. She took detailed notes.

She was given a combination of HCA and chromium, two tablets, three times daily, before meals.

Darlene liked her new 'glucose control project' and saw it as one she could master 'no sweat'.

We recommended an ECG, blood testing and an exercise treadmill test prior to the start of her exercise regime, in addition to the recommended physical examination.

She passed with flying colours and couldn't wait to get started on her walking programme. We insisted that she stay on the recommended slow start programme, as she had not been exercising routinely. She agreed and within two weeks was up to walking half an hour every day.

Re-testing of her blood revealed a fasting blood glucose level of 110 mg/dl, well within the normal range and nearly 100 points lower than when she had started. Her six-pound weight loss at the end of a month was testimony to her dedicated spirit and rapid change in lifestyle, for the better.

Darlene is a good illustration of my 60-plus age group. Frequently they will embrace the elements of the Diabetes Cure very well, because it gives them something to do!

Many retired or 'winding down' 60-year-olds are finding themselves with little planning about what to do after retirement and they become bored. Some become intimately involved with the daily soap operas, and pass their lives away from episode to episode. Many others find a renewed joy in becoming healthy.

It is this age group that is at most risk for developing the 'silent diabetes' of type 2 diabetes mellitus. As Darlene tells her friends, 'Don't be afraid of finding out you have type 2 diabetes, be afraid of *not* finding out you have it!'

Nine Steps to Better Health

There you have it, the nine-step Diabetes Cure. Although it won't put an unconditional end to type 2 diabetes in some people, it will at least reduce the amount of prescription medication that has to be taken to keep your glucose levels under control and the devastating complications of diabetes at bay.

The Diabetes Cure requires more time to take hold than prescription medicine because it is a holistic approach. You can expect to see changes in the amount of prescription medication you are taking in about two to four weeks. Even though HCA and chromium alone will most likely have positive effects on your blood glucose levels, the Diabetes Cure works best when all of the nine steps are followed.

CHAPTER FOUR

Exercise That Fights Diabetes

The Programme

How Much? At least 30 to 60 minutes of continuous exercise of any type.

How often? Every day, with no excuses.

Why? Any exercise reduces your risk of getting diabetes and increases your sensitivity to insulin if you do have diabetes. A substantial amount of research proves this. Vigorous exercises (running, hiking, bicycling) have for a long time been shown to improve insulin sensitivity. More recent research has shown that non-vigorous activity like walking and even household chores can have a very positive impact on diabetes.

The value of exercise in controlling and avoiding type 2 diabetes has been made obvious to me a number of times by the effect it has on patients. One such patient was a man named Ed, a 55-year-old teacher with a growing spare tyre around his middle and a worsening glucose problem. Tests showed him to have high glucose levels, and his physical symptoms – dizziness, sudden exhaustion, excessive urination – indicated that this silent disease was beginning to get noisy.

Believing that less medication is most often best, I tried a novel approach: an exercise prescription. I explained that for some reason, exercise helps the muscles 'suck up' glucose and utilize it. If he could walk regularly – for at least 30 minutes per day – and eat no more than he was eating, I speculated he would need no other form of intervention.

Ed liked the idea and followed this simple exercise pre-scription religiously. Every morning he got up an hour earlier and walked for more than 30 minutes. Nothing stopped him. If it rained, he wore a raincoat. If he slept through the alarm, he walked at lunchtime or in the evening. For one month, he walked diligently. His symptoms disappeared and his glucose levels dropped into the safety zone.

For Ed, exercise alone was the solution to his diabetes prob-lem. For you, it might be different. There is always one truth: Exercise is one of the most important things you can do (together with healthy eating) to keep your blood glucose levels in check and to minimize complications from diabetes. That's because exercise usually lowers blood sugar and helps your body use the food you eat in a more efficient way.

At the same time, exercise appears to help insulin work better. This important information was reinforced recently by the University of South Carolina. In studying a racially mixed group of nearly 1,500 participants, researchers found activi-ties ranging from sports and vigorous exercise to hobbies, household chores and even sleep all had a positive effect on insulin sensitivity. Reduced insulin sensitivity is one of the risk factors for diabetes as well as heart disease.

If you're overweight, then exercise – along with careful attention to nutrition and medication – can help take off the extra pounds. Losing those extra pounds can help reduce the need for medication as well as minimize diabetes complica-tions. There are more reasons why exercise should be an important part of your diabetes treatment plan. They include:

- **Benefits to the cardiovascular system.** Regular exercise, especially aerobic exercise – the kind that makes you sweat and your heart beat faster – improves the way your heart, lungs and blood vessels work. It decreases overall blood cholesterol and increases the levels of high-density lipoprotein (HDL), so-called 'good' cholesterol. When 10 marathoners were checked after they had run a race, their HDL had gone up as much as 10 per cent. One study from Brigham and Women's Hospital in Boston found men with diabetes who exercised strenuously at least once a week saw their risk of heart attack plummet 50 per cent compared to those who didn't exercise at all. More frequent exercise reduced the heart attack risk further.

- **High blood pressure control.** Exercise can help bring down high blood pressure, which is associated with stroke and heart attack. One study involving Harvard University alumni found those who were active had a 20 to 40 per cent lower chance of having high blood pressure. Studies have also shown that those who exercise regularly have a slower resting heart rate; their heart also beats more slowly when they're exercising than the hearts of people who are out of shape. A slower heartbeat and lower blood pressure translate into good news for those who want to stay healthy. Both have been shown to slow the hardening of the arteries, or even to reverse it.
- **Glucose control.** Exercise helps remove glucose from your body. So if you have type 2 diabetes, exercise may help you control your blood glucose with little or no medication. This occurs because exercise improves glucose transport and insulin sensitivity, reducing insulin resistance for 8 to 48 hours after exercising. Work with your health care team or a fitness instructor to develop your personalized exercise programme. Remember, never alter the amount of diabetes medication you take without the approval of your doctor.
- **Increased muscle strength and flexibility.** Strength-building exercise increases the muscle-to-fat ratio. When combined with stretching, the result can be a body that's stronger, has greater muscle tone and bends and flexes more easily. Weight-bearing exercise such as working out with weight machines, free weights, or just walking is especially important for women because, in combination with a healthy

diet, it can preserve bone mass and reduce the risk of osteoporosis. There is also a mind—body connection to physical activity that can lift your mood. Physical activity distracts you with an enjoyable activity that takes the immediate focus off your disease. Second, exercise increases the release of endorphins, according to a study at the University of Wisconsin, Madison. These naturally occurring brain chemicals affect how you feel. Exercise raises body temperature, which may result in brain chemical changes that help to ease anxiety. Either way, exercise is especially important for people with diabetes who may experience depression as a natural part of their disease.

- **Weight loss.** Carrying extra pounds reduces your body's sensitivity to insulin. Losing weight and keeping it in a normal range helps your body make more efficient use of the insulin it produces and may lower the risk of diabetic complications.
- **Improved body image.** How you look is directly related to how you feel. Being physically fit improves muscle tone, circulation, can help to control body weight while also helping you remain active and able to participate fully in your own life and in the care of your diabetes.

The overall benefits of all types of regular exercise for everyone, not just people with diabetes, are being recognized increasingly. Even after genetic disorders and other risk factors are considered, people who exercise regularly are much less likely to die prematurely than their sedentary counterparts, Finnish researchers report. Those who vigorously walk or jog for an average of 30 minutes six or more times a month, considered 'conditioning exercisers', had a 43 per cent lower risk of death than sedentary people, according to the study of about 8,000 sets of twins.

Even people who were occasionally active had a 29 per cent reduced risk of dying compared to more sedentary people, says Dr Urho M Kujala, an expert in sports and exercise medicine at the University of Helsinki. In the study, published in the

Journal of the American Medical Association, Kujala and colleagues observed similar risk reductions when they looked specifically at individual twin pairs in which only one twin was active. This suggests risk factors that run in families may be overcome, in part, with exercising.

Ralph Paffenbarger, MD, of Stanford University, in his research on the physical activity of 17,000 Harvard alumni, discovered those who were more active lived longer. Death rates came down steadily as the subjects burned off more and more energy by doing such things as stair climbing and walking. He noted that people who took part in rigorous sports had the lowest death rates, but not far behind were people who pushed themselves to some degree but not so hard that they broke out in a sweat. How much does a person add with exercising? According to Paffenbarger, every hour working out boosts the life span by an hour. It's almost like free time.

Start with Your Doctor

A talk with your doctor is absolutely essential before beginning any exercise programme. In addition to an assessment of your overall physical health, your doctor can determine the kinds of exercise that are right for you. This decision should take into account certain kinds of exercise you should avoid if you have diabetic complications such as degeneration of the eye (known as retinopathy), heart disease or kidney and nerve damage.

People with diabetes have a different physiologic reaction to exercise than people who don't have the disease. During aerobic exercise, those without diabetes generally have no change in blood glucose levels because the liver responds as the demand for glucose increases.

Those who have diabetes generally experience a decrease in blood glucose no matter what their pre-exercise level of blood glucose. People with diabetes should talk in detail with their health care teams about how to recognize the symptoms and avoid hypoglycaemia (low blood sugar).

Practise Random Acts of Exercise

Despite the many benefits to being physically fit and active, more than 60 per cent of adults don't get the recommended amount of activity they need; 25 per cent aren't active at all, according to one report. A busy schedule at home and at work, inclement weather, boredom with the current exercise programme, occasional illness and inertia can lead even the most dedicated person to slack off and eventually give up.

But there is good news. All exercise counts. According to a 1995 issue of the *Journal of the American Medical Association*, the accumulation of 30 minutes or more of 'moderate-intensity physical activity' every day can pay off in enhanced health. Such light workouts can even relieve anxiety, researchers have found.

Called 'random exercise', it can be a great way to extend or add to your regular exercise programme, but take note – the operative word here is *accumulated*. If you're dedicated, creative and motivated, you can get the exercise you need, improve your overall health (including blood glucose levels), feel better and maybe even live a longer, healthier life. Remember this the next time you're tempted to circle the car park a second time looking for a parking place nearer your destination.

Indeed, the benefits of random exercise can and do make a difference in your good health and to the reduction of your diabetes complications. Some possibilities include:

- **Garden work.** Mowing the grass, raking leaves, weeding or using a brush instead of a leaf-blower are all good fitness sources. When done vigorously, they have the added benefit of aerobic exercise which also increases strength and flexibility. Other possibilities – washing the car, mulching flower beds and cleaning out the gutters.
- **In the house.** Vacuum carpets and scrub floors, wash windows and clean out cupboards. Not only will you gain in fitness, you'll feel generally better about your living

environment. Other possibilities – washing down the walls or skirting-boards, waxing the floor, making the beds, shaking out the rugs, painting a room and hanging clothes out on the line.

- **At work.** Instead of ringing or e-mailing someone on the next floor or in the next building, walk over. Use a portion of your lunch hour or break time to walk in the halls or around the building. Be sure to swing your arms and walk briskly to gain maximum benefit. Other possibilities – pausing every hour to leave your desk, doing lateral stretches and twists, moving your arms and touching your toes.
- **Just for fun.** Give the golf buggy a miss and carry your own clubs on the course. Bypass the pudding and go for a walk after dinner. With your doctor's approval, pick up a light-intensity fitness activity that has a social component such as walking at the shopping precinct, bowling, dancing or skating.
- **Around town.** Instead of driving, walk from shop to shop. Carry your own shopping to the car. Facing a long delay at the doctor's surgery – spend the time walking briskly. Other possibilities – walks in the park, volunteering as a playground supervisor at a local school, or taking a water aerobics, weightlifting or yoga class.

Focus on Fitness

Random exercise is a way of augmenting a regular exercise programme, but it won't meet all your body's needs for strength, aerobic and flexibility exercise. To develop a well-rounded exercise programme, talk with your doctor or a fitness instructor about your lifestyle, exercise preferences, your overall health and your diabetes management. Work together to figure out ways to make a complete fitness plan a normal part of your life, one that fits in naturally and also meets your needs. It should include:

- **Your fitness preferences.** If you hate team sports or running, find alternatives.

- **Balance and variety.** Doing the same activity every day is boring and could mean you're overworking one muscle group to the detriment of another. Vary your programme by walking one day, taking a dance class on another day and cutting the grass with a push mower on another.
- **Your schedule.** A night owl who signs up for an early morning aerobics class is destined for failure. By creating an exercise programme that fits with your personal style, you're more likely to persist.
- **Willingness to seize opportunities.** Park your stationary bicycle or ski machine in front of the television and work out while watching the evening news. At work, walk to the toilets at the opposite end of your floor. Don't saunter in the supermarket, walk briskly. It will save you money and, in combination with carrying your own groceries, build strength.

Enjoyable exercise is the kind that falls naturally into the rhythm of your day. If your fitness activities feel dull or point-less, it's time for a change. Even though rigorous workouts normally are better for the body, most of the good results of exercise can be had through any form of activity that pushes the body somewhat.

How hard should a workout be? The answer to that question is very individual and should be based on your total health. The American College of Sports Medicine says it's best to exercise at 60 to 90 per cent of your maximum heart rate. Though, again, any activity is better than no activity. Harder activity tends to strengthen the heart more and be better for the body in general.

You can chart a proper workout level by subtracting your age from 220 to arrive at your maximum heart rate, in terms of beats per minute. Then this number should be multiplied by .60 and .90 to find the lower end and higher end of the 'target' workout range.

For instance, if you're 40, subtract your age from 220 to get your maximum heart rate – 180 beats per minute. Multiply this figure by .60 and .90. This gives you your target range: in this case a pulse rate of between 108 and 162 beats each minute. Here are target heart workout rates for people of various ages:

Target Heart Rates

Age	Target Range Beats Per Minute
25	117—175
30	114—171
35	111—166
40	108—162
45	105—157
50	102—153
55	99—148
60	96—144
65	93—139
70	90—135
75	87—130
80	84—126

To measure your heart rate during a workout, stop exercising briefly and put two fingers on your neck just below your ear, or on your wrist. Simply count the number of beats in 10 seconds and multiply that number by six to come up with the count for one minute. That's your heart rate for one minute.

Exercise Safety Guidelines

People with diabetes face special challenges when it comes to staying fit. To make the most of your fitness programme:

- Know your blood glucose level before, during and after exercise.
- Keep a quick source of carbohydrates with you at all times, such as a non-diet soft drink, dried fruit or glucose gel or tablets.
- Get an exercise buddy. If you must exercise alone, tell others where you are going and when you'll be back.

- Begin every workout with a warm-up and end it with a cool-down.
- Drink plenty of water before, during and after your workout.
- Carry diabetes identification. If possible, also carry a cellular phone and/or money for a phone call.
- In very hot or cold weather, exercise indoors – consider walking at an indoor shopping centre or purchasing exercise equipment for your home.

Nix Nicotine

You know that smoking is bad for your cardiovascular system, but did you know that as a person with diabetes it's even more urgent that you quit? Here are the facts. According to one study, epidemiologists at University College London followed 4,427 people with diabetes for 14 years. Compared with people with diabetes who had never smoked, those who had quit more for longer than 10 years still had a risk of death from all causes that was 25 per cent higher. Those who had quit more recently had a death risk 53 per cent higher. The researchers speculate that smoking plus diabetes multiplies damage to the heart and blood vessels, accounting for the unusually high risk among ex-smokers who have diabetes.

Quitting smoking reduced death risk compared with current smokers, but risks remain unusually high for ex-smokers with diabetes. And it's not just *your* smoke that can affect your health. Exposure to second-hand smoke is worst for those with diabetes as well as for those with high blood pressure, a study shows. Researchers defined 'exposure' to second-hand smoke as spending at least one hour a week near at least one smoker.

People exposed to second-hand smoke are at greater risk for atherosclerosis, the build-up of artery-narrowing plaques. In a study of 10,914 middle-aged adults compared with those free of second-hand smoke exposure, researchers found those who were exposed experienced a 20 per cent increase in atherosclerosis progression. Atherosclerosis in the people with diabetes who were exposed to second-hand smoke progressed even faster.

As a person with diabetes, you know the risks of life-threatening complications from your disease are high, so take action. With so many options to choose from when it comes to giving up your nicotine addiction – support groups, nicotine replacement systems and behaviour modification – there's never been a better time to quit. Talk with your doctor about the best way for you to manage your nicotine addiction and reduce the increased risk of diabetic complications that smoking presents.

Preventing Hypoglycaemia

The risk of hypoglycaemia (blood sugar falling to dangerously low levels) keeps many people with diabetes on the sidelines. With careful management and the help of your health care team, you can learn to exercise while also managing your disease. Here are the basics:

- Plan to exercise one to three hours after a meal when blood glucose is rising or peaking.
- Monitor your blood glucose levels and, if necessary, snack before or during exercise. How much you should snack depends on your overall health, your ability to control your blood glucose levels and the type of exercise you're doing. For instance, if you're biking, walking or golfing, a high-carbohydrate snack such as 6 fluid ounces of juice or half a plain bread roll should do the trick. On the other hand, if you're doing intense exercise such as aerobics, running or racquetball, half a meat sandwich with a cup of low-fat milk is called for.
- Know the symptoms of hypoglycaemia and respond promptly with a fast-acting sugar source such as non-diet soda or fruit juice, raisins, boiled sweets or glucose tablets.
- Know how your body responds to particular kinds of exercise by checking your blood glucose levels before, during and after exercise.

If you feel a hypoglycaemic reaction coming on, stop immediately and have 4 fluid ounces of orange juice, a non-diet soda or three glucose tablets.

Athletes and other very physically fit people may have a lower risk of hypoglycaemia because their conditioning allows them to store glucose better in their muscles. These same people may be at higher risk for post-exercise, late-onset hypoglycaemia. This condition may occur up to 24 hours after exercise and is more likely after a sudden increase in their training load or a long or high-intensity workout. The best way to prevent this type of hypoglycaemia is awareness, plus careful meal planning and adjustment of insulin during and after exercise. If you believe you fall within this category, talk with your health care team. They may suggest you meet with an exercise physiologist or other health care professional familiar with the care of athletes who have diabetes.

Exercise in a Perfect World

An intense workout should include three basic parts:

1   A 5- to 10-minute warm-up with gentle stretching. As you warm up, avoid bouncing as it can injure muscles that aren't fully ready to work.
2   20 to 30 minutes of aerobic activity that raises your heart level but doesn't cause straining or shortness of breath. A good rule of thumb for the aerobic part of your programme is that you should be able to exercise and carry on a conversation. Can't do it? Slow down a bit until you can talk comfortably.
3   5 to 10 minutes of cool-down exercises, stretching or slow-paced walking.

A good exercise programme should make you feel better, not worse. It's true, you may be a little sore the next day, but don't let minor muscle stiffness stop you. Just invest a little more time in your warm-up to stretch out those newly challenged muscles and begin again.

If you haven't exercised in a while or have been ill, you may find you can only do 5 to 10 minutes of aerobic exercise before you begin to strain. Listen to your body and slow down. Do what you can and return to your exercise programme the next day. By gradually increasing the length and intensity of your workout, you reduce the chances of injury and increase the chances that you'll stick with it.

The good news about aerobic exercise is that the benefits don't stop when you do. Your body continues to burn calories at a higher rate after your workout. The number varies, depending upon the intensity of the workout, your overall fitness level and your muscle mass. Since muscle burns many more calories than fat, the benefits can increase as your strength and muscle mass increase.

Exactly how often you should and can exercise depends on many factors, including:

- your overall health
- blood glucose levels and control
- your lifestyle
- any diabetes complications.

In general, schedule some type of fitness activities daily. You may want to alternate days when you do aerobic exercise with others when you do strength-building or flexibility-enhancing activities. Or you may want to keep the same consistent programme. The most important thing is, with your doctor's approval, get active and stay active.

Warning

For people with diabetes, sometimes exercise isn't a good idea. Talk with your doctor about your specific medical history and don't exercise if you:

- are ill or recovering from an illness
- have diabetes that has affected your nerves (called neuropathy) or your eyesight (retinopathy) and haven't discussed with your doctor the impact of exercise on these complications
- are following your care plan, but continue to have difficulty keeping your blood glucose levels under control
- have a blister or open sore on your foot
- haven't checked your blood glucose levels
- haven't eaten
- haven't talked with your health care team or a fitness instructor to develop a fitness strategy.

Foot Facts

Peripheral nerve damage can lead to a loss of sensation in the feet. This loss of sensation may mean a blister goes unnoticed and becomes infected. For the person with diabetes, this infection can be deadly. Diabetic complications are the fourth leading cause of limb amputation. Make attentive foot care an integral part of your fitness programme.

- Buy fitness shoes that fit well. The shoe should have plenty of room in the toe box, fit snugly at the heel and feel absolutely comfortable from the start. Shop for shoes in the afternoon when your foot size tends to be largest, and try them on wearing your workout socks. Spending the extra time and money necessary to get a shoe that fits perfectly is a long-term investment in your feet.
- Break in new shoes slowly by wearing them for an hour or so, then switching to another pair.
- Change shoes every day and allow them to air and dry out between wearings.
- Wear clean socks that are a combination of cotton and polyester that take moisture away from your feet at every workout. You may consider purchasing socks with extra

cushion and arch support for exercise or any time you expect to do a lot of walking.

- Inspect your feet daily for blisters, scratches, redness or tenderness, and talk with your health care team if you find any of these.
- Never walk barefoot, not even indoors.
- Wash your feet daily and dry between the toes carefully. If your doctor recommends it, apply lotion, but avoid getting it in between the toes.
- Trim your toenails with the contour of your toes. Talk with your health care team if ingrown nails are a problem.

Weight a Minute

Exercise melts body fat. Even though people who work out tend to eat more than couch potatoes, exercisers tend to have less body fat. Less fat means less of a load on the heart and circulatory system, and less need for insulin. So how does a person reap the benefits of exercise and push life to its limits without undue strain? It's not all that hard. Plenty of exercise can be had with tiny lifestyle changes. And the process can actually be fun.

Average Calories Burned in Exercise per Minute

Knowing the value in calories of some of the common forms of exercise can help to motivate you. It can also be especially important if you're trying to get fit while also losing weight. The number of calories you burn varies with the intensity of your workout and also with your size. Here are some exercise options and calorie-burning ranges per minute of exercise.

The Diabetes Cure

Activity	Weight in Pounds			
	105—115	127—137	160—170	182—192
Bicycling, indoor	11	13	15	17
Skiing cross-country, moderately hilly terrain	9	11	18	16
Running	9	10	12	13
Step aerobics	9	10	14	17
Skipping rope 100 skips/min.	9	11	15	17
Volleyball	8	9	11	12
Handball	8	9	11	12
Basketball	7	8	10	11
Stair climbing	6	7	8	9
Aerobics	6	7	8	9
Hiking 4 mph, 20-lb pack	6	7	8	9
Tennis, doubles	6	6	8	8
Square dancing	6	6	8	8
Cross-country skiing, ski machine, 11 mph	6	6	8	10
Treadmill, 13.5 mph	6	6	8	10
Downhill skiing	5	7	8	9
Gardening	5	6	7	8
Rollerskating	5	5	7	7
Calisthenics	4	5	7	8
Weight training, moderate resistance	5	6	7	8
Skating	5	5	6	7
Walking, 15 min/mile	5	6	8	9
Swimming	4	5	5	6
Sexual intercourse	4	5	5	6
Rowing	4	5	5	6
Walking, 20 min/mile	4	4	6	7
Golf	3	4	4	5
Bowling	3	3	5	5

Wellness Walking

When it comes to a fitness pay off, nothing beats walking. It's free. When done at an appropriate level of intensity, it's usually injury- and pain-free. Best of all, walking is a normal part of most people's day. To gain maximum fitness benefits from it, all you have to do is more.

Start small – 5 to 10 minutes a day is a good way to begin, especially if you haven't exercised in a long time or you're newly diagnosed with diabetes. By starting with a short increment, you can build up stamina and strength, plus the confidence you need to exercise often and well, despite your diabetes.

Walk at a pace that's invigorating and enjoyable. Move your arms briskly and inhale and exhale deeply to gain maximum benefit. Increase the length and speed of your walk slowly to allow your body to adjust and adapt. If you feel pain or strain, return to your most recent level of activity and start anew.

If goals motivate you, work over a period of weeks and months to a level of four miles per hour. If goals are a disincentive for you, just work towards gradually increasing the duration of your walks.

On the Run

Some people would rather run than walk. And it can be a great workout if you're properly warmed up, wearing appropriate shoes and accustomed to the increased demand it places on your body. Running or jogging can be hard on your joints and feet. For people with diabetes, who are already at risk for foot injuries, this can be a concern.

If you've talked to your doctor and received the all-clear to hit the jogging path, start by using a walking/running approach. Begin by warming up your muscles, especially those in your legs, and walking briskly. Once you feel warmed up, move to a jogging pace and continue until you feel winded. Then fall back to a brisk walk until you've recovered your breath; then return to jogging. Over time, you'll probably be able to jog the entire distance of your chosen route. At that point, you may want to alternate jogging days with walking days.

Where you run is up to you, but it's best to avoid concrete, which is especially punishing on the joints. Consider a local track, paths in your neighbourhood park, or just running in the grass along the pavement. Remember to listen to your body and slow down if you experience breathlessness or pain, especially in the joints.

Walker/Jogger Safety

- Wear shoes that are comfortable and allow your feet to breathe.
- When walking or jogging outdoors, dress appropriately for the weather.
- Always walk or jog facing traffic.
- If walking at night or near dusk, wear light-coloured clothing and a deflector vest, and carry a torch.
- Wear your diabetes identification and be sure to tell someone where you're going and when you expect to be back.

Strength Training

Building muscle as a part of your exercise programme has many benefits. Strong muscles help you cope with everyday activities, whether you're carrying the shopping or your baggage on a business trip, or working in your garden. Strength training is a way you can work to prevent broken bones due to osteoporosis. An extra plus to weight training is that well-toned and larger muscles burn more calories.

According to a study from the American College of Sports Medicine, weight training may lower total cholesterol by 4 per cent. More importantly, it may lower low-density lipoproteins (LDL), so-called 'bad' cholesterol, by as much as 13 per cent while boosting high-density lipoproteins (HDL) by as much as

5 per cent. This cholesterol-lowering benefit is especially important for people with diabetes, who are at a higher risk for cardiovascular disease.

Strength training usually involves repeated lifting of hand weights and/or the use of weight machines. As your strength increases, the number of lifting repetitions and/or weight of items being lifted are increased. There are many approaches to this activity. If endurance is your goal, lift moderately heavy weights 15 to 20 times, rest, then repeat. If strength plus endurance are your goals, use a weight you can lift only 8 to 12 times; rest between these 'sets' of 8 to 12 lifts.

Talk with your doctor or a fitness instructor about which approach is right for you. You may be referred to an exercise physiologist who can assist you in determining the best approach, proper lifting techniques, and appropriate weight for you to begin with.

Many people do strength training at fitness clubs where they have access to a variety of strength-building machines and the supervision of trainers or other professionals. You can gain many of the same benefits with hand weights at home. You can even create your own hand weights by filling 1-litre bottles with measured amounts of sand or water.

Smart Weight Training

Before beginning a weight-training programme, talk with your doctor. This can be especially important for people with diabetes since weight training can, in some cases, worsen diabetic complications such as retinopathy. Other safety first weight training strategies include:

- Start with 5 to 10 minutes of aerobic exercise such as walking, running or calisthenics to warm muscles.
- Breathe in when lowering weights and breathe out when lifting weights.

- Work out with a partner or a trainer who can watch for problems in your technique, spot you when you're using heavy weights, and provide encouragement.
- Either alternate your weight training day with a rest day when you focus on flexibility and endurance activities, or work muscles in the upper body on one day and the lower body the next.
- Cool down after your workout by walking, stretching or doing a slow-paced stint on a stationary bicycle or ski machine.

Staying Motivated

If you keep doing what you're doing, you'll keep getting what you're getting. That's good news when it comes to a fitness programme. Unfortunately, motivation can wane, and your well-planned schedule can get interrupted. Suddenly, a week has gone by and you've not so much as laced up your athletic shoes. Here are some tips to keep your exercise plan on track:

- **Do it often.** One way or another, everybody should work out at least three times a week. That schedule allows a day off in between exercise days to let muscle soreness go away. But this is a minimum. With a gradual build-up, it's safe to work out four or five days a week or even every day.
- **Do it long enough.** Exercise sessions should be at least 20 to 30 minutes long. But that's a minimum. Anyone who exercises more than this does his or her body more good.
- **Set goals that work for you.** Begin gradually and increase the intensity and duration of your workout over time.
  Harder isn't necessarily better. Perhaps you've heard 'no pain, no gain.' That couldn't be further from the truth. Pushing yourself to a level where you feel pain can result in injury and more than a little reluctance to continue working out. Instead, focus on the time spent exercising, not on making the exercise hard. Constancy and consistency are the keys.
- **Spice it up.** Variety is the spice of exercise life. Shift the kinds of fitness activities you do with the seasons or your family

and work schedule. Above all, make your fitness fit your day, not the other way around.

- **Work with someone.** A friend or 'fitness partner' can help keep you motivated and give you someone to compare your progress with.
- **Map your personal progress.** Track how well you've improved with a workout diary. Tracking your success, along with the success of your diabetes management, is one way to keep motivated.
- **Reward yourself.** Have you reached a milestone goal? Been especially diligent in sticking with your fitness and overall diabetes management plan? Reward yourself with a new pair of workout shoes, a class that teaches you a new fitness skill, or a change in your exercise routine that makes it easier for you to maintain your commitment.

Exercise and Pregnancy

Whether you developed gestational diabetes as a part of your pregnancy or were living with type 2 diabetes before you became pregnant, exercise plays an important part in your good health. Staying physically fit throughout your pregnancy helps moderate weight gain and increases your strength and stamina.

Contrary to prevailing beliefs, exercise may lower blood sugar during gestational diabetes, if properly supervised. Experts at the third International Workshop-Conference on Gestational Diabetes Mellitus have said that exercise – along with a proper diet and supplements such as vitamin B_6 and chromium – may help lower blood sugar. This, in turn, may reduce health risks to mother and foetus.

Monitoring your blood glucose levels is especially impor-tant during pregnancy, because the level can drop quickly as a result of exercise. Talk with your doctor or health care team about learning to maintain tight control during your preg-nancy. It may mean checking your blood glucose levels seven or more times a day, but it is an investment in your continuing good health and that of your baby.

Talk with your doctor, midwife or health care team about developing an antenatal diabetes management and fitness plan. And remember:

- If you haven't been exercising before your pregnancy, this isn't the time to start training for a marathon. Instead, go for low intensity; start slowly with walking and simple, relaxed stretching exercises.
- Follow the exercise and blood-glucose-testing regime recommended by your doctor, midwife or health care team.
- Bypass activities that require you to lie on your back after the fourth month of pregnancy, involve holding your breath, straining, or include jerky movements or quick changes of direction.
- Your blood sugar can drop quickly and unexpectedly during exercise, so be attentive to early symptoms of hypoglycaemia. If they occur, stop immediately and have fruit juice, non-diet soda or three glucose tablets.
- Drink plenty of liquids before, during and after you exercise.
- Start with warm-up and cool-down stretches and limit the intense portion of your programme to no more than 15 minutes.
- Check your heart rate frequently and keep it below 140 beats per minute.
- Be aware of your body temperature and keep it below 100°F/38°C. Don't use a hot tub or steam bath without your doctor's permission.
- Stop exercising if you feel light-headed, weak or breathless.
- Ask your midwife to show you how to feel your uterus for contractions during exercise. Contractions can be a sign you're overdoing it.

Until recently, experts thought exercise endangered the mother's health and/or foetal development. Experts now say the only pregnant women with diabetes who should avoid exercise are those who have (or have had):

- pregnancy-induced hypertension
- three or more spontaneous abortions
- cervical incompetence
- foetal distress
- placenta praevia or other placental problems.

Soothing to the Soul

Exercise is an essential part of the Diabetes Cure, but unpleasantness isn't. Feel free to try a number of different exercise activities until you find the ones that suit you best. There is more than one way to burn a calorie, and the ways you choose should be soothing to the soul as well as helpful for the body.

CHAPTER FIVE

Help Through Healthy Eating

The Programme

What? Develop a healthy diet that is probably different from the one you currently follow.

Why? Some doctors call type 2 diabetes 'hoof and mouth disease', because it is caused by not enough hoof (exercise) and too much mouth (food). Weight gain is thought to be the primary cause of type 2 diabetes. The diet recommended in these pages controls glucose levels. It also promotes a feeling of fullness. These two factors – glucose control and feeling full – will help control or prevent type 2 diabetes. They will also help you to lose weight if you need to.

How? I recommend a diet in which 50 to 60 per cent of calories come from complex carbohydrates (whole grains and high-fibre fruits and vegetables), 10 to 20 per cent from protein like beans, fish, poultry, eggs and (very little) red meat, and 20 to 30 per cent from fat, with most of that coming from monounsaturated fats like olive oil.

And don't forget to watch your salt intake.

What you eat, when you eat, and how much you eat directly affect your blood glucose levels. It's as simple as that. Many people never really give serious consideration to the types of foods they eat until they're diagnosed with diabetes. Having diabetes isn't the end of your relationship with the foods you like. Instead, it can be the beginning of a healthier relationship with food and with your body.

In truth, this meal plan is ideal for anybody who wants to eat in a healthful, nutritious way. This diet includes a wide variety of foods from all parts of the food pyramid, including new foods as well as the old favourites you love. When you follow a recommended meal plan, you minimize the risk of diabetes complications like heart disease, high blood pressure and the risk of amputation or blindness, as well as giving your body the nutrition it needs to function at its best.

The Diabetes Control and Complications Trial, a 10-year study of the connection between closely managed glucose levels and diabetes complications, found that people with diabetes can reduce their chances of complications, but never entirely eliminate them. That's a fact which should motivate you to do all you can to manage your disease carefully.

Do you have an individualized meal plan? Most people find it helpful to meet with a registered dietitian experienced in diabetes care. Think you don't need one? Think again. Your diabetes is with you for life, so we're not talking about some pre-packaged diet programme that you'll follow for a few weeks or months, then return to your old eating habits. Instead, a registered dietitian will work with you to create a customized plan.

Diet Guidelines

A dietitian will help you evaluate your:

- daily calorie intake
- daily carbohydrate intake in grams, which will allow you to keep your blood glucose levels within your target range
- daily fat gram intake, to keep you at fewer than 30 per cent of total calories
- options for breakfast, lunch, dinner and snacks
- techniques for distributing your calories throughout the day so your body always has the food it needs

- choices for adjusting your meals (or medications) for exercise and travel.

Ultimately, your meal plan is designed to teach you an eating style that's flexible and takes into account your:

- favourite foods and the ones you avoid
- lifestyle, including work schedule, family commitments and exercise
- target blood glucose levels
- ideal weight
- need to minimize the risks of diabetes complications as well as other health risks.

Anatomy of Food

Food is fuel. The right balance of foods containing needed carbohydrates, protein and fat, plus vitamins, minerals and micro-nutrients will help your body function more smoothly. Scrimp on the quality or quantity of fuel and your body pays the price. It's true for everyone, but especially true if you have diabetes. That's because the carbohydrates, protein and fat provide calories for energy and affect blood glucose levels.

No single food group provides all the nutrients and calories a healthy body needs to run smoothly, fight off infection and rebuild itself daily. It's one of the reasons the Food Pyramid is an ideal tool. When you eat from all parts of the pyramid in the recommended proportions and combinations, you provide your body with an edge in maximizing good health and mini-mizing diabetes complications.

The fact is, all foods provide the body energy in the form of calories. Food not used immediately is stored as fat. The chal-lenge for most people, but especially those with diabetes, is to strike a balance between the number of calories you take in and the number you burn. That's because being overweight, even by a small amount, makes it more difficult to keep your blood glucose levels in a normal range or to manage health-related issues such as cholesterol and triglyceride levels.

For some people with diabetes, weight management can be tricky. As they gain greater control over blood glucose levels, they may find their weight goes up. That's due, in part, because they are no longer losing calories in their urine. To maintain their weight they have to cut the number of calories they take in. Talk with your doctor, dietitian or health care team about a weight management programme that will prevent weight gain but reduce the likelihood of low blood sugar (hypoglycaemic) reactions.

Here's where using the Food Pyramid and a meal plan come in. At the very top are fats, oils and sweets. These should be eaten sparingly by anyone concerned about good health and keeping weight down. Next are dairy products and meat/protein. Small amounts of both of these are necessary, but remember that they're higher in calories than foods at the bottom of the pyramid. You should target getting the bulk of your nutrition from the last categories – vegetables, fruit and bread, cereal, rice and pasta. These are good choices because they're major sources of fibre, vitamins and minerals, and of energy that's easily accessible to the body.

Carbohydrates

Carbohydrates are your body's primary energy source. They contain 4 calories per gram, and 100 per cent of carbohydrates are metabolized into glucose. Carbohydrates have a protein-sparing effect by supplying the energy needs of the body and enabling the body to use protein for growth and repair. When broken down, carbohydrates become glucose, a simple sugar that is your body's main energy source.

Complex carbohydrates are found in some vegetables, grains, cereals and breads, while simple carbohydrates (sugars) are found in fruits, some vegetables and milk. They also show up in sweets, fizzy drinks and other sweet treats. To maximize good nutrition, go for complex carbohydrates. They provide more vitamins, minerals and dietary fibre per calorie than simple sugars. Be wary of such processed foods as white bread. Refining removes much of the nutritive value. Opt, instead, for whole grains.

In general, get 50 to 60 per cent of your total calories from carbohydrates.

Despite what you may have heard, how quickly glucose gets into your body depends not only on the type of carbohydrates you've eaten, but on several factors, including:

- how much is eaten
- how foods are cooked or prepared
- other foods eaten at the same time.

This is true even though simple sugars and complex carbohydrates are digested at the same rate, because their overall digestion may be slowed by fat present in the food.

Proteins

Proteins are used by the body for tissue growth and repair. Like carbohydrates, they provide 4 calories per gram. Even though protein is an important part of a balanced diet, don't go overboard. Unless you choose very lean cuts of meat or eat a vegetarian diet, many sources of protein can be high in fat calories.

Protein sources include meat, fish, poultry and dairy products. When eaten in combination, additional protein sources include rice, beans, eggs and grains.

As a group, proteins provide your body with essential amino acids.

Plan to get 10 to 20 per cent of your total calories from protein sources.

Cutting Back on Fat

- It's all in the cooking. Skip the frying pan and grill; roast, steam, or poach meat instead.
- Spray away. Use a cooking spray to prevent sticking during the cooking process.
- Fixate on fin and fowl. Overall, seafood and skinless poultry is lower in fat than red meat, so make steaks and chops an occasional treat and focus on seafood and poultry.
- Go lean. Select lean cuts of meat. If tenderness is a concern, pound it or create all-natural tenderizers using citrus juices and spices.
- Be size-wise. Cut back on your portion to about 3 ounces – that's a serving about the size of a deck of cards.
- Go Eastern. Cook as Asians do, using very small amounts of finely sliced or chopped meat as flavouring to complement generous servings of vegetables. Then stir fry in a non-stick pan using a non-stick spray or a tiny amount of monounsaturated oil.
- Skim and trim. Trim excess fat from meat. Chill soups, prepared broth or casseroles and skim off the fat that rises to the top before reheating.

Fats

Fat has gained a bad reputation in the last few years. The fact is, fats are necessary in limited quantities because they provide the fatty acids which are necessary for life. Fats are stored and used as energy reserves. The real issue is how much and which types of fat you eat.

In his Report on Nutrition and Health, US Surgeon General C Everett Koop took aim squarely at high consumption of foods laden with fat and cholesterol as two elements that make a major contribution to the high incidence of heart disease and other cardio-vascular ailments. That's because excess calories from any source are stored as body fat and, unfortunately for most people, the fat storage capacity of the human body is unlimited.

Fats provide 9 calories per gram, making them the most concentrated calorie source.

To make the most of good nutrition, limit your fat intake to 30 per cent of your total calories. That's a good idea for everyone, but especially important for people with diabetes. If you have diabetes you are at increased risk for premature atherosclerotic vascular disease, and that can increase your risk of heart attack and stroke.

There's also a connection between the amount of body fat you have and insulin resistance. Having too much fat, especially on the upper part of your body, makes it difficult for your body to use insulin. Being overweight puts a strain on your pancreas which, in turn, makes it harder for the pancreas to make the insulin you need.

Cholesterol is the type of fat most often associated with diabetic complications. Usually, your body makes this substance on its own. It's essential in the making and repair of cell membranes and plays a role in the manufacture of such essential hormones as oestrogen and testosterone, but it is one of the main culprits in clogging arteries and causing heart disease.

Saturated fat is found in animal and some vegetable products. To avoid these you have to be a smart shopper and read food labels and the ingredients of foods you buy. Saturated fats show up most often in fast foods and such processed foods as commercially baked goods, cake mixes, biscuits, crackers, coffee creamers and snack foods. Even some oat bran cereals which gained popularity for their ability to lower cholesterol contain large amounts of saturated fat. While these foods may contain no cholesterol, they do contain substances such as palm-kernel oil, which has been shown to raise cholesterol.

From saturated fat the liver produces a cholesterol-rich bile which digests fat. Since saturated fats are such complex molecules, more bile has to be produced to digest them than for unsaturated fat. The result? Much of the extra cholesterol ends up circulating in your bloodstream as saturated fat.

And there's more. Saturated fat reduces the number of low-density lipoproteins (LDL so-called 'bad' cholesterol) receptors in the liver. These receptors pull LDL cholesterol from the bloodstream. If they aren't there to perform that function, the LDL continues to build.

Believe it or not, there are even some fats which, when eaten sparingly, are good for you. These include polyunsaturated fats and monounsaturated fats, both of which contain fewer hydrogen atoms and don't stay in your bloodstream the way their saturated cousins do. Unsaturated fats require less bile to digest. They also bind with cholesterol and pass into the intestines and on into the bloodstream. In the *Archives of Internal Medicine*, Dr Grace Goldsmith published results showing an unsaturated fat diet leads to as much as 25 per cent more bile being eliminated from the body than a saturated-fat diet. The result? Lower cholesterol levels in the bloodstream.

The ultimate goal is to get no more than 30 per cent of your total calories from fat, with no more than 10 per cent coming from saturated fats.

A Word of Caution

Remember that 'fat free' on the label does not mean calorie-free, or that you can eat all you want. For instance, 1 ounce of sour cream has 62 calories and 6 grams of fat, while the reduced-calorie version has 45 calories and 2 grams of fat. Many fat-free biscuits make up for the absent fat content by increasing the sugar content. The only way to be sure of exactly how much fat and how many calories you're getting is to read the label carefully. Recommended serving size should get equal attention. Eating just two low-fat biscuits can ultimately save you calories and fat grams. But some people are seduced by the low-fat label and may eat twice as many of the low-fat version, ultimately consuming more calories than if they'd eaten the regular version.

Fat Cutters

Interested in cutting back on fat, but not sure how to begin? Here are some basic changes you can make that will help you cut back on fat:

- Cut fat in recipes by reducing the amount you add. If a dish calls for 4 tablespoons of oil, add two and make up the rest with water. When using oils choose corn, safflower or other polyunsaturated types.
- Use skimmed milk instead of whole.
- Replace sour cream with plain, low-fat yoghurt.
- Use only egg whites or egg substitutes in cooking.
- Substitute fat-free frozen yoghurt, sherbet or sorbet for ice cream.
- Substitute a fat-free artificial topping for whipped cream.
- Select low-fat or fat-free margarine instead of butter. In particular, look for margarines that have vegetable oil as the first ingredient, not partially hydrogenated oil.
- Opt for lower-fat cheese such as mozzarella, made with part-skimmed milk, ricotta and low-fat cottage cheese.
- Create your own lunch meats using skinless baked chicken or turkey.
- Select fat-free mayonnaise over traditional or homemade mayonnaise.

Common Sources of Saturated Fats

butter	lard
meat fat	palm oil
cocoa butter (chocolate)	soft cheese (e.g. Philadelphia-brand)
bacon and bacon fat	sour cream

Common Sources of Polyunsaturated Fats (Healthier Fat Choices)

corn oil	cottonseed oil
sunflower oil	safflower oil
soybean oil	mayonnaise
salad dressings	margarine

Common Sources of Monosaturated Fats (Healthiest Fats)

olive oil	canola oil
avocados	peanut butter
nuts	

Friendly Fibre

Fibre is the indigestible portion of plant food that passes through the intestines nearly intact. There are two main types of fibre – soluble and insoluble. Soluble fibre slows the rate of nutrient absorption from the stomach and intestines and may be helpful in lowering blood sugar and cholesterol levels. Examples include: oat bran, fruits, vegetables, beans, peas, nuts and seeds. Insoluble fibre speeds the passage of food through the digestive system. It increases the bulk of the stool, aiding in the prevention or treatment of constipation, which can be a problem for some people with diabetes. Examples include whole grains and wheat bran.

The average diet contains only about 10 to 20 grams of dietary fibre per day. If you have diabetes it's a good idea to work towards gradually increasing your fibre intake to 25 to 50 grams per day. Increasing the fibre in your diet too rapidly may result in excess wind, cramping, bloating and diarrhoea. Adequate fluid intake – six to eight glasses per day (a total of about 2 to 3 pints) – is also important when fibre is increased.

Salt Is Everywhere

Sodium – salt – is everywhere and the average person eats 5,000 milligrams per day. That's not hard to imagine when you consider that just 1 teaspoon of salt contains 2,000 milligrams of sodium.

If you have diabetes, cutting back on sodium is a long-term investment in your good health. People with diabetes are at an increased risk for high blood pressure. The general recommendation for sodium intake is 1,100—3,000 milligrams per day.

Hypertensive patients should restrict sodium to 2,000 milligrams per day or lower, as ordered by their doctors.

Sodium causes your fat cells to retain fluid, expanding them to as much as eight times their size. This additional weight, sometimes called water weight, is an extra burden on your body, especially your heart. Some evidence suggests that too much salt over a long time can lead to kidney malfunction, which forces the blood pressure higher.

High blood pressure and even salt itself has been shown to speed the development of atherosclerosis. Not only does the increased pressure contribute to plaque deposits, but it causes smooth muscle cells in the arteries to contract, become stiff and enlarged. The larger the artery muscles become, the smaller the lumen – the area where blood flows through the muscles. Taking in less sodium smoothes these muscles cells, causing them to relax to a smaller size so the lumen area is larger.

Outside your kitchen and without a lot of label reading, it can be difficult to identify foods high in sodium, which can appear in the most unlikely places. For instance, a McDonald's milk shake has a higher sodium content than their Quarter Pounder. It's even found in products such as AlkaSeltzer, sugar-free drinks and penicillin. Other likely suspects include:

- salad dressings
- sandwich meats and processed meats
- tinned soups and vegetables
- fast foods
- some cheeses
- snack foods such as crisps and crackers.

To reduce sodium intake, stop using salt in cooking and at the table, and avoid processed foods. Concerned food won't taste as appetizing? Experiment with combining fresh and dried herbs and spices as well as adding a dash of herbed vinegar or fresh citrus juice. Other salt-reducing tips include:

- Avoid processed foods.
- Use salt substitutes.
- Eat more fruit.

Sweeteners

Sugar substitute possibilities are everywhere. The pink stuff, the blue stuff – whatever your choice, they can help people with diabetes satisfy their sweet tooth without falling off the food-plan wagon.

Technically, most sweeteners are classified as nutritive or non-nutritive. All of the following sweeteners are approved for use by persons with diabetes.

Nutritive Sweeteners

FRUCTOSE

Fructose, found in fruit and honey, is one of the most common naturally-occurring sugars. It is not associated with rapidly increasing or high levels of blood sugar in well-controlled diabetes. It is one to one-and-a-half times as sweet as real sugar and contains the same number of calories. Products sweetened with fructose should be calculated into the diet as carbohydrate calories (either starch or fruit) if more than 20 calories are used at one time.

SUGAR ALCOHOLS

Sugar alcohols include sorbitol, mannitol and xylitol. They contain 4 calories per gram and are absorbed slowly and metabolized like fructose. If you have enough insulin, the fructose will be stored as glucose in the liver. If there is not enough insulin, the fructose can cause a rise in plasma glucose. Sugar alcohols can cause diarrhoea.

ASPARTAME

Best known as Nutrasweet, aspartame is a combination of two amino acids – aspartic acid and phenylalanine. Aspartame is

180 times sweeter than sucrose, so only a small amount is needed to get the sweetness you want. It contains 4 calories per gram but does not affect glycaemia (blood sugar levels).

Non-Nutritive Sweeteners

SACCHARIN

Three hundred times sweeter than sucrose, the body doesn't metabolize saccharin. The kidneys excrete it unchanged so it has no calories. Although some studies have shown saccharin causes bladder cancer in animals, it's considered safe at human levels of consumption.

Smart Shopper

Feel baffled by all the food claims you see in adverts and on food packaging? You are not alone. Most consumers ignore these claims or guess at their true meaning. Let your diabetes motivate you to sharpen your food-shopping techniques. Once you and your dietitian have set daily goals for calories, fat, saturated fat, cholesterol and carbohydrate, as well as fibre and sodium, use food label information to help you shop smart. Reading food labels can be time-consuming, but it's worth it.

Start by looking at the nutritional information listing that is a part of the packaging of nearly all foods sold in the UK. It tells you the basics such as the serving size, the number of calories per serving, and the calories from fat. Serving size, in particular, deserves your attention. Many foods such as sweets or snacks may appear to have modest amounts of fat and calories, until you learn that technically, only a tiny amount is calculated as a serving size.

Next comes the actual number of calories, grams of fat, cholesterol, sodium, carbohydrate and protein in 100 grams and/or a single serving of the product.

Health Claims

Food manufacturers can only make health claims on food labels that are supported by scientific research. Some of these include the relationship between:

- calcium and osteoporosis
- fibre-containing grain products, fruits and vegetables and cancer
- fruits and vegetables and cancer
- fruits, vegetables and grain products that contain fibre – particularly soluble fibre – and the risk of coronary heart disease
- fat and cancer
- saturated fat and cholesterol and coronary artery disease
- sodium and hypertension
- folate and neural tube defects.

Other Claims

Manufacturers also make claims about the nutritional value of products which can be confusing. Here are some of those claims and what they really mean:

- Calorie-free. Has fewer than 5 calories per serving or other designated amount.
- Low-calorie. Has 40 calories or fewer per serving.
- Light or 'lite'. Has one-third fewer calories or half the fat than the food(s) it is being compared with, usually the full-calorie version of the same food.
- Less and reduced (as in fat or sugar). Is at least 25 per cent lower in calories or other ingredients compared to the full-calorie or non-reduced version.
- Cholesterol-free. Contains fewer than 2 milligrams of cholesterol and 2 grams or less of saturated fat per serving.

- Low-cholesterol. Contains 20 milligrams or less of cholesterol and 2 grams or less of saturated fat per serving.
- Low-fat. Must have 3 grams or less of fat per serving to fall in this category. This can be confusing, because although vegetable oils contain no cholesterol, they are 100 per cent fat. Vegetable oils are still preferable to butter or lard because they have less saturated fat. A tablespoon of vegetable oil still has about 14 grams of fat and the same 126 calories found in butter or lard.
- Fat-free. Describes a food with less than 0.5 grams of fat per serving.
- Low in saturated fat. Describes a food with 1 gram or less of saturated fat per serving and not more than 15 per cent of its calories from saturated fat.
- Low-sodium. Contains 140 milligrams or less of sodium per serving and per 100 grams of food. Remember, sodium is also found in monosodium glutamate (MSG), sodium bicarbonate (baking soda), and sodium nitrate, and occurs naturally in some foods.
- Very low sodium. Contains 35 milligrams or less of sodium per serving and per 100 grams of food.
- Sodium-free or salt-free. Has fewer than 5 milligrams of sodium per serving.
- Light in salt. Has 50 per cent less sodium than the normal version.
- Sugar-free. This describes items with fewer than 0.5 grams of sugar per serving.

More Useful Terms

Be aware that 'dietetic' has no standard meaning and only indicates something has been changed or replaced. It could contain less sugar, less salt, less fat or less cholesterol than the normal version of the same product. 'Natural' also has no specific meaning except for meat and poultry products, in which case it means no chemical preservatives, hormones or similar substances have been added. On other food labels, 'natural' is not restricted to any particular meaning. 'Fresh' can only be used to describe raw food that has not been frozen, heat-processed or preserved in some other way.

Going Shopping

When shopping, aim for foods rich in vitamins, minerals and fibre and low in fat and cholesterol. Read the ingredients listing on all food labels. Ingredients are listed in the order that they appear in the product: those that make up the largest percentage of the product are listed first, and those that make up smaller percentages of the total are further down the list. Here are some general food-buying guidelines:

- Bread. Look for low-fat varieties with whole grains as the first label ingredient: A Swedish study has shown that sourdough bread may be a good choice. That's because its fermentation process produces a bread that is less likely than other breads to trigger sharp rises in blood sugar.
- Cereal. Pick cereals listing whole grains first on the label and containing 3 or more grams of dietary fibre, 1 gram or less of fat per serving and 5 grams or less of sugar per serving.
- Crackers/snacks. Again, let whole grains be your guide and look for those with 2 grams or less of fat per serving. Try to keep the sodium content under 400 milligrams per serving.
- Rice, pasta and whole grains. Choose converted, brown or wild rice of any type. Look for unfilled fresh or dried pasta, preferably made with wholegrain flours. Avoid pastas containing eggs and fat.
- Frozen desserts. Choose ones with 3 grams or less of fat per 4-ounce serving, especially low-fat frozen yoghurt or low- or nonfat ice cream. Frozen fruit juice bars with fewer than 70 calories per bar are another option. Avoid those made with cream of coconut, coconut milk or coconut oil, which are high in saturated fat.

- Milk. Choose skimmed or low-fat milk, buttermilk made from skimmed milk and low-fat and nonfat yoghurt that is artificially sweetened. Also consider acidophilus milk as an aid to digestion.
- Cheese. Look for skimmed-milk and reduced-fat cheeses with 6 grams of fat per ounce or less.
- Red meat. Choose lower fat grades of meat such as lean body parts, such as beef (sirloin), pork (tenderloin) and lamb (leg).
- Luncheon meats. Look for lean or 95 per cent fat-free meats with 30 to 55 calories per ounce and 3 grams of fat or less per ounce.
- Poultry. Breast meat is the leanest. Removing the skin before cooking cuts fat by 50 to 75 per cent and cholesterol by 12 per cent. Beware of turkey or chicken luncheon meat, hot dogs and bacon as they can be high in fat.
- Seafood. Fresh seafood is low in fat when it's roasted, grilled or poached, not fried. Choose tinned fish packed in water or with the oil rinsed off. Look for low-sodium products.
- Vegetables. Opt for fresh and frozen vegetables, which are the most nutritious per bite. Drain and rinse tinned vegetables to reduce sodium.
- Fruit/fruit juice. Choose fresh, frozen or dried fruit, preferably without added sugar. Look for 100 per cent pure fruit juice.
- Margarine/oils. Choose brands containing 1 gram or less of saturated fat per serving (usually one tablespoon).
- Soups. Choose low-sodium and/or reduced-fat varieties. For cream soups, use nonfat or low-fat milk or water when preparing.

Away from Home

Whether eating out is a special treat or a routine part of your day, keep the following in mind:

- Ring ahead. Be prepared for eating in a restaurant you've never been to before by ringing in advance to ask about menu options so you can plan your food for the day.
- Ask questions. Not sure what's in a particular dish or how it's prepared? Ask. If the waiter can't answer, then ask for the manager.
- Develop a good eye. Restaurant portions can sometimes be huge. Help yourself by cultivating an eye for portions that fit in with your meal plan. Some restaurants allow you to order smaller portions at reduced prices. If larger portions are served, ask for the extra to be put in a 'takeaway' box before you begin to eat.
- Go light. Ask that no butter or salt be used in preparing your meal.
- On the side, please. Request that sauces, gravy, salad dressings, sour cream and butter be served on the side or left out altogether. Although this may seem like deprivation, try it for two weeks and you will see that restaurants frequently overload food with fatty additives.
- Choose your preparation method. Opt for roasted, baked, poached or grilled meats and fish rather than fried. If food is breaded, peel off the outer coating to get rid of the extra fat.
- Ask for substitutions. Most restaurants will allow you to substitute low-fat cottage cheese, jacket potato or vegetables for chips.

I recommend eating salad with low-fat dressing with lunch and dinner, since lettuce is both filling and a good source of fibre.

Time, Please

Having diabetes doesn't mean you have to give up having an occasional beer, or glass of wine or spirits, but it does mean you'll have to plan alcohol into your diet.

Alcohol provides more calories than carbohydrates or protein (7 calories per gram), but no essential nutrients. Broken down at a constant rate in the liver by specific enzymes, your body burns about 1 ounce of alcohol per hour. Metabolizing alcohol does not require insulin, but it can cause high or low blood sugar, depending on the circumstances surrounding its use.

Consumed with a meal, it has a tendency to add calories and possibly increase the blood glucose level. On an empty stomach it can cause significant hypoglycaemia in people on insulin. That's because the alcohol inhibits gluco-neogenesis – the formation of sugar from protein and other sources. The hypoglycaemic effect may persist for as long as 12 hours after your last drink.

To be on the safe side, use alcohol in moderation. The symptoms of alcohol intoxication and hypoglycaemia are similar, so it's easy to mistake low blood glucose for intoxication and delay treatment. Another concern is that while you're drinking you may overeat or forget to check your blood glucose level.

Beer, lager, wine or spirits can fit in a diabetes food management plan, but only if you remember the following:

- Alcohol should be consumed with a meal or snack, not on an empty stomach.
- The quantity of alcohol should be limited to two units once or twice a week (1 unit equals half a pint of beer, one glass of wine or one measure of spirits).
- Excessive alcohol consumption may impair your ability to recognize hypoglycaemic reactions.
- Alcoholic beverages should be substituted for the appropriate food exchanges (that is, foods containing an equal amount of carbohydrates).
- Avoid sweet wines, liqueurs, sherry and sweetened cocktails.
- When using such caloric mixers as fruit juice, the calories in the mixer must be counted.
- In cooking, alcohol evaporates, leaving only the flavour, so using it as an ingredient in cooking is fine.

- If you're having difficulty controlling your diabetes, miss out the alcohol altogether.

The Vegetarian Way

Not everyone eats from all areas of the food pyramid. Vegans eat a diet that's based on vegetables, bean and bean products and such carbohydrates as potatoes, pasta and breads. Ovo-lacto vegetarians also eat such animal products as milk, cheese, yoghurt and eggs.

Vegetarians who happen to have diabetes can enjoy both a healthy life and their chosen eating style, but only if they follow a few basic ground rules.

A *complete* protein is a food which contains all essential amino acids. Animal proteins are classified as complete proteins; plant foods do not contain all of the essential amino acids in proper quantities and are regarded as low-quality or incomplete proteins. By combining foods from any two of the following groups of vegetarian protein sources, vegetarians can ensure they get essential amino acids:

- Soybeans
- Dried beans, peas and lentils
- Grains, including oats, wheat, barley, rice and sweetcorn
- Nuts and seeds
- Milk products.

These foods complement one another when eaten over the course of a day. Confirmed complementary proteins include grains and legumes, seeds and legumes, and grains with milk products. Typical dishes might include red beans and rice, nut loaf with sweetcorn, or cheese quiche with split pea soup.

Eating widely from the many alternative protein substitutes is one way to be sure you get the nutrition you need as well as the necessary amino acids. That's where proteins such as tofu, cheese, eggs and nut butters come in.

Vegans need to include a source of vitamin B_{12} in their diets, for example cereal or soy milk fortified with vitamin B_{12}. In addition, if their exposure to sunshine is limited, a vitamin D supplement may also be needed.

When You're Ill

Whether it's a cold or a serious illness, a person with diabetes needs to take special measures when unwell. Start by working with your doctor to develop a sick-day plan. That plan should serve as your guide for helping you get well and also include specific information about when to call the doctor.

In general, check your blood glucose levels according to the schedule outlined on your sick-day plan. If possible, eat normal meals. If you're too ill to eat, sip small amounts of juice every 10 to 15 minutes.

Diarrhoea or vomiting requires further special attention, because you will be losing sodium, potassium and fluids which need to be replaced. Sodium can be replaced by adding salt to foods and by eating tinned soup, broth and salted crackers. Foods high in potassium include orange juice, grapefruit juice and vegetable juices, bananas, oranges, potatoes and dried fruits. Many salt substitutes are also high in potassium.

CHAPTER SIX

The Vitamins and Minerals Edge

The Programme

What? Taking daily vitamins and mineral supplementation will add to the vitamins and minerals already found in a diet that is appropriate for a person with type 2 diabetes.

Why? A growing body of research shows that vitamin and mineral supplementation can improve the way your body responds to insulin. This research shows that vitamins with antioxidant properties lessen the impact of type 2 diabetes by 'mopping up' free radical oxidants which occur as part of the disease process of diabetes and can damage nerves and blood vessels.

How? Make certain you are getting enough of the vitamins and minerals discussed in this chapter.

'Should I take more vitamins now that I have diabetes?'

'How much vitamin C is enough?'

'Are there enough vitamins in my food?'

These questions are asked of all medical doctors who treat people with diabetes. The answer the patient receives will be different, depending upon the bias of his doctor. Many doctors, for instance, recommend no vitamin supplementation at all. They feel that we get more vitamins and minerals in our daily diet than we need and shouldn't bother with extra pills.

Others take the 'better safe than sorry' approach. Although they don't believe in vitamin and mineral supplementation, they don't disbelieve in it either. To their way of thinking, a little extra supplementation won't hurt.

Others, like myself, heartily endorse the use of vitamin and mineral supplements. We have seen the positive effects of these supplements on patients with diabetes and know that they have virtually no side-effects.

For many years, Western medicine has been opposed to the taking of supplements. When I went to medical school, the prevailing thought was that 'health nuts' were the only ones who took supplements. We were encouraged not to recommend supplements to our patients. Some of our lecturers even felt that it amounted to quackery to do so.

Now things have changed. A growing body of medical research is showing that vitamins and minerals have a place in medicine, especially in the successful treatment of type 2 diabetes. This chapter focuses on that research and provides a clear path for you to follow in giving your body the essential nutrients it needs to battle diabetes.

Outside Influence

Vitamins and minerals, along with glucose, phytochemicals and protein as well as amino, fatty, linoleic and linolenic acids, are the essential nutrients your body doesn't make but which you need in order to live a long, healthy life. They are the foundation of good nutrition, because in their absence your body can't use the food you eat.

People with diabetes know that, along with exercise and following their doctor's recommendations regarding medication, getting the essential vitamins and minerals is one way to manage their disease successfully. Certain vitamins and minerals appear to pack a double punch for people with diabetes, because they also may have a role in reducing the risk of diabetic complications and may help to reduce the need for medication.

The use of vitamins, minerals and other naturally occurring substances in foods is the new frontier in the treatment of diabetes as well as many other diseases. With in-depth research and continued study, scientists are unlocking the secrets these substances hold in disease prevention and management.

If you're one of the many people who think vitamins and minerals are the fuel that runs your body, think again. In fact, it's the *calories* in food that give you energy. Vitamins and minerals release that energy by allowing your body to digest, absorb and metabolize nutrients. In essence, they are the key which, when turned, runs the engine of nutrition.

Vitamins

Thirteen vitamins essential to good health have been identified. These fall into two categories – water-soluble and fat-soluble. The former group – vitamin C and the eight B vitamins – are found in the watery portions of food; they are absorbed directly into the bloodstream, circulating freely throughout the body. Fat-soluble vitamins – A, D, E and K – can only be absorbed in the presence of fat. Fat-soluble vitamins enter the bloodstream via the intestinal wall. They are stored in fat storage sites in the body.

For the most part, the best way to maximize the nutritional potential of water-soluble vitamins is to eat as much of them as possible when they're fresh or frozen. Improper storage can lead to water-soluble vitamin loss. That's why it's important to store such foods, including citrus fruits, green vegetables, peppers, tomatoes, berries and potatoes, dairy products, liver and other offal, clams and oysters, eggs, fish, chicken, pork, whole-wheat products, brown rice and oats, carefully and in covered containers.

Avoid overcooking, as this can deplete the vitamin content. For instance, peas are an excellent source of thiamin (vitamin B_1), but if you overcook them they can lose as much as 40 per cent of their thiamin content. Milk is a good source of riboflavin (B_2), but this vitamin is sensitive to ultraviolet and fluorescent light. So milk stored in a clear container at the shop may have riboflavin depletion even before you get it home.

Minerals

Minerals are the most basic element of nutrients. They can't be destroyed, unlike vitamins which can be lost easily in cooking and storage. But that doesn't mean every mineral you take in is easily and immediately available to your body. Take spinach, for instance. It's packed with calcium, but most of the calcium is not available because it is bound to oxalic acid and that blocks absorption.

There are at least 16 minerals identified as being essential to good health. Some are categorized as major minerals, others trace minerals. The difference is in the amount the body needs, not in their importance to nutrition.

Medical research is only just beginning to uncover the delicate interrelationship between vitamins, minerals and disease management and prevention. What we do know is that it's not enough simply to eat according to the Food Pyramid or even to take a daily vitamin supplement.

About Antioxidants

Everyone's talking about antioxidants. That's because preliminary epidemiological studies seem to show a relationship between these substances and disease prevention. The theory is that vitamins C and E, plus beta carotene and other carotenoids, as well as minerals selenium, copper and manganese, may be instrumental in supporting good health through their actions in the body.

Antioxidants appear to protect cells from damaging oxygen-free radicals or oxidants. Oxygen-free radicals are released in the body as the result of the breakdown of food into fuel. These free radicals appear to damage cells in much the same way that oxygen causes some metals to rust or oxidize.

This cell damage may be connected to a plethora of health problems, including heart disease, cancer, cataracts and a weakened immune system, to name a few. In addition to the free radicals created as a natural part of the body's processes, most people also come into contact with free radicals through such external forces as tobacco smoke, pollution, and even rancid dietary fats. Remember this the next time you eat fast food chips that taste bitter, are exposed to second-hand smoke, or are tempted to leave your car idling while you run an errand.

Researchers at the Royal Brompton Hospital in London have linked abnormally high levels of blood glucose to the production of free radicals. They believe these free radicals may trigger a cascade of undesirable chemical reactions, leading to what they call 'oxidative stress'. The result? For people with diabetes, the potential for complications such as heart disease, blindness, kidney disease and nerve damage.

Free radicals seem to wreak havoc on the body in a variety of ways. They appear to cripple the cell membrane, preventing it from taking in nutrients, oxygen and water and getting rid of waste products; they may even kill cells. They also may attack cells' genetic material, making the cells more likely to replicate and grow abnormally – essentially the disease process that makes cancer so deadly.

Fortunately, the body has its own way of dealing with free radicals. To prevent their damaging effects, antioxidants step in. Some appear to control the free radicals; others seem to transform them into less damaging compounds; still others are believed to repair damaged cells.

Not all antioxidants are alike. In fact, they aren't interchangeable. Each has a unique job to do in nutrition and health. For instance, E may be most effective at protecting the arteries against heart disease and the prostate against cancers; carotenoids appear to guard the retina against degeneration.

Using an assay that measures isoprostane levels in urine, researchers at the University of Pennsylvania Medical Center have found a way to gauge oxidative damage caused by free radicals. This measuring system may become an integral tool in clearly understanding the action of antioxidants in protecting the body from free radical damage. Researchers continue to study the action of antioxidants in an effort to determine how they can be better harnessed to enhance disease fighting and prevention.

Phytochemical Facts

Phytochemicals are the new kids on the nutrition block. Classified as non-nutrient parts of food, the work of phytochemicals is still being studied. What we do know is that they give plants their colour and protect them from free radicals as well as viruses, extremes of heat and cold, and rough handling. Don't let the non-nutrient label fool you. While they have no calories and appear not to be crucial for normal function, researchers believe they are able to extend to people some of the same protection they give plants.

In animal studies, it has been shown mice fed a diet of such food components as starch, protein, vitamins and minerals were more likely to develop liver tumours than those fed whole foods. Phytochemicals are believed to act at three different points to prevent cancer – they seem to prevent cancer-causing substances from forming, boost the activity of blocking agent enzymes, and suppress precancerous activity.

Phytochemical Profiles

To date, researchers have identified only a handful of what are believed to be hundreds of phytochemicals. The existence and action of the remaining ones are still to be revealed. Some phytochemicals may hunt down free radicals, preventing LDL cholesterol from oxidizing and reducing the risk of heart disease as well as the risk of cancer. Here are just a few of the known phytochemicals, their food sources, and believed roles in disease prevention.

Ellagic Acid
Common in grapes, strawberries, raspberries and walnuts, this phytochemical appears to remove cancer-causing agents from the bloodstream.

Flavonoids
A family of chemicals found in apples, red wine and tea, flavonoids seem to play a role in preventing cancer and heart disease. They may reduce the risk of cancer by blocking the hormones that create cell changes and by suppressing changes in malignant cells.

Indoles and Isothiocyanates
Cruciferous vegetables – Brussels sprouts, cauliflower and cabbage – are especially good sources of antioxidant phytochemicals.

Isoflavones
Soybeans contain this phytochemical, along with several other antioxidant substances, including phytic acid, protease inhibitors, saponins, phytoserols and phenolic acid.

Lignans
Commonly found in oil pressed from flax seed, lignans is a powerful antioxidant that also contains omega-3 fatty acids which may protect against heart disease and colon cancer.

Monoterpenes
Commonly found in citrus fruits and caraway seeds, this phytochemical can prevent oxidation and keep potent carcinogens from taking hold.

Organosulphur Compounds
Found in garlic, onions and other foods from the *Allium* genus, this phytochemical has been found in studies to have a possible impact on reducing the risk of colon and stomach cancer.

Phenol

Tea, especially green tea, is a good source of phenol. It seems to activate enzymes that eliminate oxidants.

Why a Pill Isn't Enough

Vitamins, minerals and supplements are all the rage these days. From your local chemist's generic brand to pricey designer elixirs, you can buy a dizzying combination of substances, all of which claim to be indispensable in enhancing overall health, boosting energy or resolving a particular deficiency.

The fact is, no pill or supplement can substitute for a balanced diet. A rich and varied diet containing fruits, vegetables, grains, dairy products and protein also includes fibre and phytochemicals essential for good health. A pill can't come close to supplying all these in the right combination or form, but when used to supplement the diet can be very helpful to the body.

Balance is another concern. Good nutrition is based on taking in appropriate amounts of vitamins and minerals to act on the calories you eat and burn. When you add large amounts of any one vitamin or mineral, you run the risk of disrupting this delicate balance.

There's also a risk of taking in, via a vitamin or supplement, a quantity that can damage your health. For instance, vitamin A is necessary for good vision, but too much can damage your liver. Taking large amounts of zinc because you heard it can augment your immunity could put you at risk of a copper deficiency.

How much of any one vitamin or mineral is needed is set by Government and EU standards. One such set of standards has been known as the recommended dietary allowances (RDAs) of vitamins and minerals. RDAs are guidelines for safe and adequate levels of nutrition, though are not optimal or even minimal requirements. This leaves room for personal needs based on age, lifestyle and health. Keep this in mind when you're studying food labels or considering the purchase of a vitamin or mineral supplement.

The RDAs aren't dietary commandments. They don't dictate every substance each person needs for good health because, in part, medical researchers are only now beginning to explore the intricate and individual relationships between vitamins, minerals, trace minerals and other substances supplied in our food. Rather, RDAs are estimates based on the calculated dietary needs of healthy people whose food intake varies from day to day. If you have a chronic disease – high blood pressure, irritable bowel syndrome or diabetes – you may need to augment your diet, either with specific foods or supplements, in order to make up for vitamins and minerals lost as a result of your disease or the medications you take.

Finally, RDAs don't take into account the changing nutritional needs that occur over a lifetime. For instance:

- Menstruating girls and women may need extra iron to compensate for blood loss during monthly flow.
- Women of childbearing years need extra folate before and during the first few weeks of pregnancy to reduce the risk of neural tube defects.
- Pregnant or breastfeeding women may need extra calcium, iron and vitamin B_6, plus folate to compensate for increased demands on their bodies.
- Adolescents and older men and women need extra calcium to build bone and prevent osteoporosis, *especially* if you are a person with diabetes.
- Older adults, who are less active and less inclined to eat a balanced diet and whose bodies may not absorb nutrition efficiently, may need a multi-vitamin.

Whether you should be taking a multiple vitamin or specific supplements is a decision for you and your doctor or dietitian to make based on your personal health, your lifestyle, and your age.

Diabetes: The Vitamin and Mineral Connection

Researchers increasingly are looking to vitamins and minerals as potential tools for treating people with diabetes. These substances may hold the key for long-term good health, managing complications and, possibly, even prevention.

Heart Disease

Cardiovascular disease is an ever-present concern for people with diabetes who are at greater risk. According to the American Diabetes Association, adults with diabetes have heart disease death rates two to four times higher than adults without diabetes. At the same time, 60 to 65 per cent of people with diabetes have high blood pressure; their risk of stroke is two to four times higher than people without diabetes.

University of Texas Southwestern Medical Center researchers described recent experiments in which people with diabetes and people without the disease were given 1,200 international units (IU) of vitamin E daily. After eight weeks, the process that leads to a build-up of fatty deposits in the arteries was slowed.

Therapy with vitamin C and other antioxidants shows promise in helping to reduce the incidence of cardiovascular disease in some people with diabetes. The study, conducted at Brigham and Women's Hospital and Harvard Medical School, found short-term infusion of vitamin C improved blood vessel function in 10 patients with diabetes, but not in the 10 who did not have diabetes.

The researchers studied the endothelium – the smooth layer of cells lining the inside of blood vessels which helps them to dilate and to contract properly. Previous research had indicated that the endothelium could be damaged by oxygen-derived free radicals; antioxidants such as vitamin C are believed to destroy these free radicals.

Findings regarding vitamin E's antioxidant properties in preventing hardening of the arteries also are encouraging. In a study of middle-aged people who took vitamin E supplements, there was markedly less build-up of fatty plaque in the neck arteries than among test subjects who did not take supplements. Even among those who did not take supplements, the higher the person's dietary intake of vitamin E, the less arterial clogging.

The Nurses' Health Study, an ongoing investigation of many preventative health measures involving 70,000 female nurses, has shown a 50 per cent drop in the risk of heart disease among participants who consume high amounts of folic acid and B_6. Folic acid and B_6 are believed to interfere with the action of homocysteine, a chemical that worsens atherosclerosis. This study may have relevance for reducing the risk of atherosclerosis, even though researchers endorsed good weight management and stopping smoking as the keys to the management of heart disease overall.

A study from the University of Maryland School of Medicine has found consuming high doses of vitamins C and E before a high-fat meal may counter the fat's ill-effects. Researchers believe these vitamins somehow keep triglycerides (a type of blood fat) from preventing the release of nitric oxide by the blood vessels. Nitric oxide allows the vessels to relax and widen, facilitating healthy blood flow to the heart.

Their study found that before a fatty meal, blood vessels could widen by 21 per cent to accommodate the sudden rush of blood, but four hours later could only widen by 8 per cent. When study subjects consumed 1 gram of vitamin C and 800 IU of E before eating a high-fat meal, the amount by which their arteries could widen only fell to 17 per cent four hours later.

Researchers believe these vitamins, which are antioxidants, squelch free radical action. Previous research has found that high cholesterol levels interact with oxygen in the body to produce an excess of oxygen-free radicals. These are known to interfere with nitric oxide and cause blood vessels to be less pliable. This impairment of blood vessels is a recognized risk factor for atherosclerosis.

Smoking Solutions

That smoking is linked to many types of cancer, high blood pressure and cardiovascular disease is well documented. These associations are just some of the reasons people with diabetes are strongly encouraged not to smoke.

Researchers at the University of Freiburg in Germany and also at Boston University Medical Center have linked the taking of liquid vitamin C by smokers with impaired arteries to improved blood flow.

Vitamin C is believed to clear free radicals left behind by smoking from the body. The free radicals damage endothelial cells, causing increased deposits of low-density lipoprotein (LDL), so-called 'bad' cholesterol, and narrowing of the arteries. These two conditions can lead to heart disease and heart attack.

Calcium and Men

The positive effects of calcium in protecting women from osteoporosis or 'soft bones' is well known, but for men with diabetes, even a small shortage of calcium could increase their risk for the same condition. Osteoporosis is a leading cause of broken bones, debilitation and death among the elderly, 18.4 per cent of whom (people 65 and older) have diabetes.

Researchers at Athens University Evangelismos Hospital tested 30 men and women with diabetes for bone loss in the spine and thigh. These measurements were then compared with people who did not have diabetes. They found that men with diabetes lost significantly more bone than men without the disease over the course of the study. Women with and without diabetes lost bone at the same rate.

None of the women in the study was on oestrogen replacement therapy, which helps to preserve bone strength. Diabetes experts hypothesize that because older women continue to produce small amounts of fatty tissue even after the meno-pause, and many women with diabetes are overweight, they may be protected from osteoporosis.

This study is believed to refute the myth that people with diabetes have stronger bones because of their greater attention to good nutrition.

Vitamin Know How

Each vitamin and mineral plays a complex role in nutrition and overall good health. Use this guide to help you under-stand the jobs of each and its particular importance to people with diabetes.

Vitamin A – The Eyesight Vitamin

Vitamin A is especially important for people with diabetes who are at increased risk for diabetic retinopathy, the leading cause of blindness among adults. Essential for vision and a strong immune system, vitamin A also helps to build and maintain strong bones. Found in liver, fish liver oils, whole and fortified milk and eggs, this vitamin can be synthesized from carotenoids, naturally occurring in orange fruits and some vegetables.

RDAs
Men   1,000 micrograms (mcg)
Women   800 mcg

Possible Negative Effects of Higher than Recommended Intake
Liver damage as well as headache, vomiting, blurred vision, hair loss, flaking skin, bone and muscle pain, spontaneous abortion and birth defects.

Beta-carotene – Antioxidant

Converted as needed to make vitamin A, good sources of this vitamin include spinach, collard greens, broccoli, carrots, peppers, sweet potatoes, apricots and peaches. Through its antioxidant action, beta-carotene protects cells from free-radical damage.

RDAs

Men	6,000 mcg
Women	4,800 mcg

B₁/Thiamin – Energy Vitamin

Essential for people with diabetes, who may experience nerve damage as a complication of their disease. Thiamin acts as a coenzyme to get energy from carbohydrates, protein and fat; it also converts excess carbohydrates to fat and is essential for the initiation of nerve impulses. Good sources include yeast, lean pork, offal, legumes, seeds, nuts and unrefined cereal.

RDAs

Men aged 19—50	1.5 milligrams (mg)
Men aged 51 and over	1.2 mg
Women aged 19—50	1.1 mg
Women aged 51 and over	1 mg

B₂/Riboflavin – Hormone Regulator

A must for the overall good health of people with diabetes. Riboflavin is an active element in the release of energy from carbohydrates, fat and protein. It is found in milk yoghurt, cottage cheese, meat, leafy green vegetables and whole-grain or enriched breads and cereals. Essential for growth and development, riboflavin also regulates certain hormones and aids in the development of red blood cells.

RDAs

Men aged 19—50	1.7 mg
Men aged 51 and over	1.4 mg
Women aged 19—50	1.3 mg
Women aged 51 and over	1.2 mg

B₃/Niacin – Fat Synthesizer

Since people with diabetes have a higher-than-average incidence of cardiovascular disease, getting enough niacin is
important in minimizing the risk of heart disease. As a coenzyme that releases energy from carbohydrates, niacin is found
in abundance in meats, especially liver, as well as fish,
legumes, nuts and whole-grain and enriched breads and cereals. Also found in coffee and tea, niacin is needed to break
down fats and proteins and to synthesize fats and certain hormones as well as aid in red blood cell formation.

RDAs

Men aged 19—50	19 mg
Men aged 51 and over	15 mg
Women aged 19—50	15 mg
Women aged 51 and over	13 mg

Possible Negative Effects of Higher than Recommended Intake
Associated with gastrointestinal complaints, liver damage,
aggravation of ulcers.

B₄/Biotin – Carbon Dioxide Carrier

People with diabetes require biotin to help their bodies metabolize fat, since fat can contribute to coronary artery disease,
one of the complications of diabetes. A coenzyme in energy
metabolism which also carries carbon dioxide, biotin is found
in liver, egg yolks, soybeans, yeast, cereals, legumes and nuts.

Used to synthesize fatty acids, biotin also assists in energy metabolism and the breakdown of certain amino acids.

RDAs
None established, but 30—100 mcg is believed to be a safe and adequate daily intake for adults.

B$_5$/Pantothenic Acid – Diverse Producer

A must-have for those with diabetes, who need to be exceptionally attentive to overall good health, especially when eating a diet that's low in fat. While helping to metabolize fats, carbohydrates and proteins, pantothenic acid, found in meats, whole-grain cereals, legumes, milk, vegetables and fruits, also assists in the production of fats, cholesterol, bile, vitamin D, red blood cells, neurotransmitters and steroid hormones.

RDAs
None established, but 4—7 mg is believed to be a safe and adequate daily intake for adults.

Possible Negative Effects of Higher than Recommended Intake
Associated with diarrhoea, a condition for which many persons with diabetes are already at risk.

B$_6$ – Multi-faceted Helpmate

B$_6$'s role in reducing the risk of heart disease makes it essential for people with diabetes. Fulfilling many essential nutritional roles, B$_6$ helps to synthesize amino acids, nucleic acids and red blood cells as well as to aid in the metabolism of proteins and urea. In combination with other B vitamins, B$_6$, found in chicken, fish, kidney, liver, pork, eggs, brown rice, soybeans, oats, whole-wheat products, peanuts and walnuts, may play a role in reducing the risk of heart disease.

RDAs

Men	2 mg
Women	1.6 mg

Possible Negative Effects of Higher than Recommended Intake
Nerve damage, a complication for which many people with diabetes are already at risk; also extreme sensitivity to light.

Folate – Cell-Growth Vitamin

Another B vitamin and a good source of support to aid people with diabetes in their fight against heart disease. By helping to convert vitamin B_{12} to a coenzyme, folate assists in the synthesis of DNA in rapidly growing cells. In combination with folate and vitamins B_6 and B_{12}, folate may reduce the risk of heart disease. Commonly found in yeast, offal, green and leafy vegetables and some fruits, a lack of folate is linked to birth defects such as spina bifida. Some specialists recommend that women of childbearing age get 400 mcg of folate daily.

RDAs

Men	200 mcg
Women	180 mcg

Possible Negative Effects of Higher than Recommended Intake
Can mask symptoms of vitamin B_{12} deficiency and pernicious anaemia.

Vitamin B_{12} – Another Cell Builder

At greater risk for nerve damage, a complication of diabetes, B_{12} supports the overall good health of those with this disease. Essential to the maintenance and replacement of cells and the nervous system, B_{12} plays a role in the metabolism of carbohydrates and fats and the synthesis of proteins and certain amino acids. Found primarily in animal products such as offal, clams and oysters, milk products, other seafoods and eggs, B_{12}, in combination with other B vitamins, may play a role in reducing heart disease.

RDAs

Men	2 mcg
Women	2 mcg

Possible Negative Effects of Higher than Recommended Intake
According to the Harvard Medical School Health Group Publication on Vitamins and Minerals, no toxicity is reported for B_{12} at intakes of up to 100 mcg.

Vitamin C – The Citrus Vitamin

Even for those who can maintain good control of their diabetes, C is significant for its cell-building and stress-management capacities. More than 80 per cent of vitamin C in the typical diet comes from citrus fruits, green vegetables, peppers, tomatoes, berries and potatoes. Its actions are diverse. It acts as an extracellular antioxidant to protect cells, regenerates oxidized vitamin E and protects iron and promotes its absorption. Also commonly found in fortified breakfast cereals, it is essential to the formation of collagen and cell walls and helps the body deal with stress and illness.

RDAs

Men	60 mg
Women	60 mg
Smokers	Up to 100 mg

Possible Negative Effects of Higher than Recommended Intake
May interfere with urine testing for diabetes and for blood in the stool; not advisable for people with a history of kidney stones.

Vitamin D – The Sunshine Vitamin

Some studies have shown men who have diabetes are at increased risk for osteoporosis; vitamin D may help to protect them by aiding in the formation and strengthening of bones by making calcium and phosphorus more available. Vitamin D also stimulates the retention of these minerals by the body. Commonly manufactured by the body via exposure to the sun, vitamin D is found in fish liver oils, fatty fish, egg yolks and fortified milk.

RDAs

Men aged 25 and over	5 mcg
Women aged 25 and over	5 mcg

Possible Negative Effects of Higher than Recommended Intake
Can cause damage to the cardiovascular system and kidneys, two complications that people with diabetes already face.

Vitamin E – Cell Protector

People with diabetes often experience dry skin; getting plenty of E can help alleviate this condition. Essential for normal cell structure and the formation of red blood cells, vitamin E is widespread in food. The best sources include vegetable oils, wheat germ, nuts and green leafy vegetables. Also an antioxidant, vitamin E seems to protect the lungs and other tissues from damage by free radicals and pollutants.

RDAs

Men	10 mg alpha TE
Women	8 mg alpha TE

Possible Negative Effects of Higher than Recommended Intake
Impaired immune function and excess bleeding as well as gastrointestinal discomfort, a complication for which people with

diabetes are already at risk; not recommended for people taking anticoagulants.

Vitamin K – Clot Creator

Diabetic kidney disease is the leading cause of people having to go on kidney dialysis. Since vitamin K is one of the vitamins that supports kidney function, it is fundamental for good health. Of the 13 proteins essential for blood clotting, vitamin K contains at least six. Also playing a role in the synthesis of proteins found in plasma, bone and the kidneys, vitamin K is widespread in food, including green leafy vegetables.

RDAs

Men aged 25 and over	80 mcg
Women aged 25 and over	65 mcg

Possible Negative Effects of Higher than Recommended Intake
May cause a form of anaemia, high levels of bilirubin in the blood, and jaundice in newborns.

Minerals

Calcium – Bone Builder

Getting plenty of calcium is vital for people with diabetes because of this mineral's important role in blood pressure regulation. It's also basic to bone formation and maintenance, and may play a role in reducing the precancerous changes in the lining of the colon. Also found in green leafy vegetables, tofu, sardines and salmon with bones and calcium-fortified orange juice, calcium aids in blood clotting and nerve-impulse transmission.

RDAs

Men   800 mg

Women   800 mg

Possible Negative Effects of Higher than Recommended Intake
May inhibit the absorption of iron and zinc; very large amounts can lead to excess calcium in the blood and kidney damage, a concern for people with diabetes who are already at risk of kidney disease.

Chromium – Metabolizer

Working with insulin to help the cells absorb glucose and produce energy, chromium also supports the metabolism of carbohydrates, protein and lipids. It is found in processed meats, whole-grain products, brewer's yeast, calves' liver and wheat germ.

RDAs
Up to 1,000 mcg per day is recommended for people with diabetes.

Possible Negative Effects of Higher than Recommended Intake
Too much chromium can result in stomach pain, cramps and diarrhoea.

Copper – Energy Resource

Copper is a part of an enzyme system that consumes oxygen or oxygen-free radicals, which have been linked to cardiovascular disease, a complication of diabetes. Needed for energy and wound healing, copper is found in shellfish, nuts, seeds, legumes, offal, bran and germ portions of grains as well as most grains, meats, mushrooms, tomatoes, bananas, grapes and potatoes.

RDAs

Men	1.5—3 mg
Women	1.5—3 mg

Possible Negative Effects of Higher than Recommended Intake
Too much copper is associated with stomach pain, nausea, vomiting, diarrhoea, coma, collapsed blood vessels and liver failure.

Fluoride – Tooth Protector

Dental problems are common among people with diabetes and fluoride can help with good dental hygiene. A beneficial mineral that protects against tooth decay, fluoride is available in tea, salmon and sardines with bones, and fluoridated water.

RDAs

Men	1.5—4 mg
Women	1.5—4 mg

Possible Negative Effects of Higher than Recommended Intake
Can impair bone health, a concern for men with diabetes who are at higher than average risk of osteoporosis. Also linked to kidney damage, another concern for people with diabetes who are already at risk of kidney disease; also impaired muscle and nerve function, and linked to mottling of the teeth, nausea, diarrhoea, chest pain, itching and vomiting.

Iodine – Regulator

At increased risk for nerve damage, people with diabetes need this mineral to maintain nerve and muscle function. It's also an integral part of two essential thyroid hormones and helps to regulate growth and reproduction. Available in sea water and iodized salt, it also assists in the regulation of the synthesis of blood cells, body temperature and the rate at which energy is released from nutrients.

RDAs

Men	150 mcg
Women	150 mcg

Possible Negative Effects of Higher than Recommended Intake
Enlargement of the thyroid gland.

Iron – Oxygen Transporter

Essential for overall good health, iron is basic for people with diabetes. By working with proteins to release energy in metabolism, iron helps to carry oxygen in the bloodstream and muscles. Iron also is involved in making amino acids, hormones and neurotransmitters and is found in meat, poultry, fish, fortified cereals and some fruits and vegetables.

RDAs

Men aged 19 and over	10 mg
Women aged 19—50	15 mg
Women aged 51 and over	10 mg

Possible Negative Effects of Higher than Recommended Intake
Can cause nausea, vomiting, diarrhoea, rapid heartbeat, weak pulse, dizziness, shock, confusion and even death.

Magnesium – Metabolizer

Another mineral that's crucial in the total health of people with diabetes, magnesium works in hundreds of chemical reactions in the body to metabolize food and transmit messages between cells. It's widespread in all unprocessed foods; good sources are nuts, legumes, whole grains, green vegetables and bananas.

RDAs

Men	350 mg
Women	280 mg

Possible Negative Effects of Higher than Recommended Intake
Associated with nausea, vomiting and low blood pressure. Very high concentrations can cause respiratory impairment, coma and heart attack.

Phosphorus – Another Bone-Builder

Phosphorus has many functions that are fundamental to good health for people with diabetes, including bone and cell growth, maintaining the body's acid/base balance, releasing energy and working with the transport of lipids in the blood and nutrients both in and out of cells. It's found in milk, meats, poultry and fish. Also found in legumes, fruits, alcoholic beverages, coffee, tea and some fizzy drinks.

RDAs

Men	1,200 mg
Women	800 mg

Possible Negative Effects of Higher than Recommended Intake
Irritation of the gastrointestinal tract, cramps, headache and, in extreme cases, liver and kidney damage.

Potassium – Transmitter

For people with diabetes, potassium acts as a transmitter of nerve impulses and muscle contraction, and maintains normal blood pressure. The richest sources of potassium include unprocessed foods, especially fruits, many vegetables, legumes and fresh meats.

RDAs

Men	2,000 mg
Women	2,000 mg

Possible Negative Effects of Higher than Recommended Intake
May cause muscle weakness, vomiting and cardiac arrest.

Selenium – Antioxidant Helper

Selenium has many functions that are fundamental to good health for people with diabetes. Commonly found in seafood, offal, cereals and grains, selenium is a component of an enzyme with antioxidant properties and may free vitamin E for other tasks.

RDAs

Men	70 mcg
Women	55 mcg

Possible Negative Effects of Higher than Recommended Intake
In very high doses can lead to nausea, diarrhoea, hair loss, nail changes, irritability, fatigue and nervous system disorders.

Sodium – Volume Regulator

Basic to nerve transmission, sodium plays a role in maintaining the good health of tissues at risk of damage from diabetes. Primarily a regulator of fluid volume, salt helps to maintain acid/base balance in the body. Salt is integral to muscle contraction. Table salt is the best source.

RDAs

Men	500 mg
Women	500 mg

Possible Negative Effects of Higher than Recommended Intake
When taken over long periods, can lead to fluid retention and high blood pressure; the latter is a special concern for people with diabetes, who are already at risk of high blood pressure.

Zinc – Metabolizer

Indispensable for people with diabetes because their disease may impair wound healing, a process zinc supports. A part of more than 100 enzymes, zinc plays a role in growth and development, immune function, blood clotting, wound healing and sperm production. Available in meat, liver, eggs and seafood, zinc also plays a role in the production of the active form of vitamin A in visual pigments.

RDAs

Men	15 mg
Women	12 mg

Possible Negative Effects of Higher than Recommended Intake
Linked to a lowering of copper levels. As a result, red blood cells shrink and immunity may be impaired. Gastrointestinal irritation and vomiting also may occur.

CHAPTER SEVEN

De-Stressing for Diabetes Care

The Programme

What?: Use the stress-reduction techniques recommended in this chapter before they are needed and when stress sneaks up on you.

How Often?: I recommend at least two concentrated sessions of stress reduction per day. That can mean performing your 30 minutes of exercise in the morning and a 30-minute session of meditation in the evening. Stress reduction can be practised at every opportunity during the day. If you feel yourself becoming stressed or angry, stop yourself, clear your mind by concentrating on your breathing, and count slowly to 10. There is a list of immediate interventions for stress reduction suggested in this chapter.

Why?: Stress-reduction intervention like meditation and even taking quiet 'time outs' has repeatedly been proven to lower blood pressure in test subjects. This is true even if the test subjects are heavy drinkers, obese, consume excessive salt or have diabetes. The lower your blood pressure (within reason) the less your chances of having complications of diabetes like heart disease or stroke. An immediate benefit of stress reduction is that it makes your life more pleasant, immediately.

A 'time out' allows you to refocus on what is really important – namely, taking care of yourself.

I had an opportunity to see stress in action when a patient with diabetes, whom I will call Tim, came in to an examining room at my clinic. He was tight with anger and didn't hesitate to tell me why.

'First I had one of those days at work,' he said, scowling at me. 'Then the traffic across the bridge was so heavy that I was late for our appointment. But I guess it didn't matter because when I got here I had to wait for nearly 30 minutes in the waiting room.'

His speech had been explosive. When he finished he was panting from the increased heart rate his anger had caused. Beads of sweat covered his forehead, and his face was red from the high blood pressure that had resulted from the outburst. Before I could say anything, he looked at me with a blank stare and turned pale. Sitting down from dizziness, he was suddenly exhausted.

'I have to quit that,' he said of his display of anger. 'Diabetes and stress don't mix.'

What Tim said was true. Although stress is a natural part of life, if you have diabetes, learning to avoid, anticipate and manage stress are key in keeping your insulin levels under control, and essential for living a normal, healthy life, despite your diabetes.

Some stress – taking a cranky two-year-old to the creche when you're already late to work while driving in rush-hour traffic during a thunderstorm – is obvious. But what about the anxious feelings you may get when you think about all your in-laws coming for dinner or that important presentation at work? Things don't have to be going badly for you to feel stressed. After all, the excitement of travelling to your child's university graduation or starting a new job can create the same physiologic reactions as negative stresses.

Many Forms of Stress

What people think is stressful is a matter of perception. Ask hospitalized patients what is currently stressful for them and they're likely to say 'Nothing, I'm not doing anything.' In fact, just being in the hospital, having their regular schedules disrupted and with the spectre of a new diagnosis or treatment hanging over them are three very stressful factors. Though the primary school teacher may feel perfectly at ease amidst 20 seven-year-olds, an options trader used to the heat of a trading session might break out in a cold sweat in a classroom with that same group of children.

Even socioeconomic status can produce stress. A recent study from the University of Michigan at Ann Arbor tracked 1,100 adults over age 29 and measured the impact of a variety of indicators on personal health. They determined that the longer or more common episodes of hardship in a person's life, the more severe the impact on health.

The skills you need for coping with different stressful situations are as varied as the stresses. You might be terrified at the idea of surfing, while the experienced surfer might be horrified at preparing a gourmet dinner for 10. Stress at work, in particular, can vary widely.

Richard Lazarus, a stress researcher at the University of California at Berkeley, makes a distinction between two kinds of coping. He describes emotion-focused coping, including talking out a problem or concern with a spouse or friends, to be most appropriate for such issues as anxiety, pointless worrying or a bad mood.

Problem-focused coping is more goal-orientated and includes intervention tactics that alter the path of threatening events or at least create potential strategies for alteration.

The trick is to develop your stress-management skills while distinguishing between the things you can control and those you can't, then taking appropriate action. Or at the very least, accept that it's out of your hands and move forward from there.

Stress and Diabetes

Stress causes real physiological and biochemical changes in the body, sometimes called the 'fight or flight' response. In its simplest form, it works like this – you feel frightened and your body responds by releasing chemicals that make a quick escape possible. You can also call it the survival instinct, because it's the basic response that allows you to step quickly out of the way of a speeding car.

Even unexpected stresses that seem merely disruptive at the time – a flat tyre or a surprise visit from a friend – can lead to jumps in glucose levels. Remember this the next time you test your glucose levels and find they're up, even though you've been dedicated to managing your diabetes well.

Most stresses aren't caused by matters of life or death. They're caused by matters of convenience or inconvenience, of personal preference, of deadlines and of life's many demands and unmet expectations. When your body feels one of these, it activates the sympathetic nervous system and increases circulating levels of catecholamines, glucocorticoids and growth hormones. These give your muscles and nerves the extra energy they need to respond by raising the level of blood glucose and ketones. Stress hormones also stimulate lipolysis (fat breakdown). This release sends free fatty acids and glycerol into circulation to be oxidized into ketones.

In the person with diabetes, stress causes elevated blood glucose and ketones. Research has found that people with diabetes differ from those without the disease in that they are particularly sensitive to these hormone responses. In their most severe reactions, that unplanned jump in glucose levels can even lead to ketoacidosis, which can cause diabetic coma.

Stress may also masquerade as hypoglycaemia because many of the same responses occur in both situations. The result? Patients may misinterpret the symptoms of hypoglycaemia – palpitations and shakiness – with psychological stress. The fear of experiencing hypoglycaemia itself can also disrupt your daily life and add to your stress.

Illness is another form of stress that many people with diabetes underestimate. Whether it's a bad cold, recovering from surgery or an ongoing battle with hay fever, the stress of illness can cause your body to release hormones that help you fight the disease but also make it difficult to control your blood glucose levels. That's why blood glucose monitoring is essential when you're unwell. The best way to cope with the stress of illness is to talk with your doctor in advance to create a 'sick day plan' that helps you manage your diabetes during illness.

Sometimes a rise in glucose and difficulty in getting it back in control signal an undiagnosed illness or infection.

A Matter of Choice

You can't always control your hormones, but you can control the choices you make that *do* control your hormones. That's central to the idea that your type 2 diabetes is, for the most part, under your control and potentially curable.

Unfortunately, when some people feel stressed they try to relax by drinking large amounts of alcohol, bingeing on junk food, cheating on sleep or engaging in careless behaviour such as missing meals. These behaviours may be benign for most people, but they can be deadly for those with diabetes. Each of these has the potential for increasing blood glucose levels, which are directly associated with a variety of dangerous complications, including depression and impotence, which can lead to further stress.

So what can you do? Take responsibility. Take time to assess and identify the stresses in your life. Then figure out which ones you can eliminate, and get to work on eliminating them. For the stresses you have to live with, create strategies that allow you to live with them and still manage your diabetes well.

That's where exercise can come in. When planned as a part of your regular diabetes management and with the support of your doctor, exercise is the single most effective way to reduce stress. Make exercise an integral part of your day. Do not wait until you 'find the time'; *make* the time, even if you exercise in only a passive way. For example, do your own garden work or even wash your car as ways of getting passive exercise. Exercise will fit most naturally in your life when it's a routine part of your day and involves activities you enjoy. (See Chapter 4 to learn more about how exercise fits in with living well with diabetes.)

Mental Strategy for Stress

Some people just seem to be more stressed than others. This has to do, in part, with personality type and coping styles. Research has shown that 'Type A' people, those who get angry or hostile when the pressure is on, tend to have higher glucose levels when they're under stress than so-called 'Type B' people who are less inclined to let life's bumps and hassles affect them.

Though it's probably true that you can't change your personality, you can change the way you respond to stress. You must learn to problem-solve instead of simply reacting, and be proactive about stress management. Here are some techniques you can use the next time you feel stress mounting. Start by asking yourself:

- Is this life-threatening?
- What's the worst possible thing that could happen?
- Will this really matter next week? Next month? Next year?
- What can I realistically do now to alter how I feel?
- What's in it for me if I get angry, upset, anxious, whatever?
- What can I realistically do in the future to prevent this from becoming a problem again?

Many high-stress people set themselves up for a severe stress reaction because they enjoy the 'high' they get when their bodies engage in 'fight or flight'. They're impatient people who demand perfection from themselves and others. They overbook their schedules so they're always running late. Or worse yet, they don't keep a daybook of commitments so they're always forgetting something important. They fail to schedule time to manage their diabetes, to eat healthy meals, check their glucose levels, exercise and relax. A Duke University study contends that people with diabetes who are under stress are five times more likely to die before the age of 50 than their calmer, more trusting counterparts.

What you say and do are choices that affect not only how you feel, the quality of your life and your diabetes, but also other people around you. In *Love and Survival*, Dean Ornish writes about the choices we have in how we treat others:

> *I am finding that I have a choice in every moment to keep my heart open or closed, to live in love or in fear. More than any specific practice, I have found that maintaining this awareness of choice is the most important factor in keeping an open heart, for every action, every thought, every moment contains the potential for bringing us closer to either intimacy and healing or isolation and suffering. The direction is not inherent in the actions themselves but rather in the intentionality and motivation behind the actions.*
>
> *The same hand that caresses also can kill. The same voice that soothes also can attack.*

Make a Change for the Better

- Count to 10 before speaking up or taking action.
- Keep a calendar of your commitments to others and yourself.
- Allow an extra 10 to 15 minutes for every activity, especially those that depend on other people or require transportation time.
- Learn to say 'no' to unreasonable or inconvenient requests, or requests that in some way interfere with your diabetes management.
- Learn to say 'yes, please' when help is offered.

- Follow your diabetes management plan regarding diet, exercise and glucose monitoring.
- Keep a diary of your blood sugar levels, food intake, exercise and other activities which you can use to be proactive in stress management.
- Plan quiet time to rest and recoup.

Taking Time for Yourself

'I don't have time to relax.'

It's a common excuse and sometimes a valid one. Your work, family and community all make demands, but if you're not feeling your best or if your blood glucose levels are above normal, you're actually cheating yourself and putting yourself at risk of life-threatening complications. Let the Diabetes Cure be your personal wake-up call to put your disease at the top of your priority list.

In *Anatomy of an Illness*, Norman Cousins wrote about his own encounter with disease. Through his experiences Cousins discovered that hope, faith, love and an abiding will to live, combined with a sense of purpose and a willingness to laugh (especially at himself), altered the quality and length of his life.

His personal experiences were later supported by findings from Carnegie Mellon University, in which a study of 400 people found psychological factors influenced the odds of fighting infection. Further, a Stanford University study determined that people with positive mental outlooks as well as strong social supports tend to take better care of themselves. They seek medical attention regularly, they eat nutritionally wholesome foods and bypass self-destructive behaviours like smoking, drug use and alcohol abuse.

All of these behaviours are used to cope with daily stressors. You can replace any behaviour with a different one if you have the will to do so. Realizing we are all creatures of habit, we need only make new habits.

The Art of Relaxation

Learning to relax is an art which must be cultivated and practised. Relaxation can take many forms. For some people a quiet evening in front of the television is restorative. For others a round of golf, an afternoon in the garden or praying quietly is refreshing.

Just conscious breathing can bring on relaxation and such benefits as lowered blood pressure, a slowed heart rate and calm thinking. According to Robert Fried, a psychophysiologist and author of *The Breath Connection*, the way most people breathe – rapidly and only expanding the upper part of the chest – leads to a chronic state of low-grade hyperventilation. The solution? Relearn how to breathe in a way that takes advantage of the diaphragm, the muscles separating your chest and abdomen. Consciously relaxing the abdominal muscles on the inward breath lets the diaphragm drop, opening up more space for the lungs. You'll know you have it right when your stomach bulges slightly as you inhale.

For people with diabetes, planning relaxation into the day is smart management and demands full and undivided attention.

Simple Relaxation Techniques

Breathe with Intention
Spending 10 to 20 minutes observing your breathing is one of the best ways to restore body, mind and spirit.

Begin by loosening tight clothing and removing your glasses. Sit or lie in a comfortable position with your arms and legs uncrossed.

For a few moments, just follow your breathing with your mind's eye as it moves evenly in and evenly out. Gradually, begin to lengthen the exhalation. So if you inhale to a count of four, exhale to a count of six, then eight, then ten and so on. You may also wish to pause for a count of one or two at the bottom and top of the exhalation. If you feel breathless, return to normal breathing, then start again. If your mind wanders, try keeping focused on your breath while repeating to yourself: 'I am breathing in. I am breathing out.'

Change Your Thinking

Some people say 'As the mind goes, so goes the body.' If you are inclined to see the dark side of life or are plagued by negative thoughts, those stresses are likely to show up in your body in the form of high blood glucose levels as well as headache, backache, stomach ache, sleepless nights and a host of other medical problems.

Changing life-long thinking patterns demands diligence and practice. Pay attention to the negative thoughts you have and identify the activities and people who bring them out. Spend 20 minutes every day focusing on the positives in your life. Mentally revisit your favourite holiday spot. Focus your attention on a beautiful flower or photograph. Make a list of things that make you smile, then post that list on the fridge, the bathroom mirror or your computer. Keep a journal in which you intentionally list 10 positive things each day. If negative thoughts emerge, acknowledge them, then refocus your attention on the positive. The notion that there is power in positive thinking may seem clichéd, but medical science shows that it can lead to a tremendous improvement in your health and happiness.

Relax Progressively

This technique teaches you to recognize when parts of your body are tense and provides the awareness needed to relax them.

Begin by lying or sitting in a relaxed position. Focus on your right leg and tense all the muscles in it. Hold for 5 seconds and then release. Do the same with the left leg. Then move to the arms, abdomen, chest, neck and head. Breathe slowly and deeply throughout, and observe the contrast between tensed and relaxed muscles.

Laugh Regularly and with Gusto
Have you ever noticed how much better you feel after a good laugh? Laughter facilitates the release of endorphins, brain chemicals that lift mood. Seek out opportunities that make you laugh, whether it's lunch with a friend, reading a funny book or taking in a good comedy film. Norman Cousins, during his illness, found Marx Brothers' films lifted his spirits and actually allowed him to cut back on pain medication.

Anger, Denial and Depression

Depression, anger and denial can all be elements or ingredients of stress. Together they can make it difficult for you to stick with your diabetes management plan.

Denial is a common reaction to hearing the diabetes diagnosis, a change in your treatment plan or the occurrence of a complication. Denial is a handy tool because it allows you to deal with bad news. It is easier to deny you have a medical problem, since this response acts as a shield from the chronic nature of diabetes and its many complications. If you are pretending there is nothing wrong, those around you will do the same, and that doesn't help in the long run.

Anger is another element that can contribute to stress. The spectre of complications or reactions, repeated trips to see your doctor and the frustration of sticking with your care plan can leave you feeling angry and helpless.

Time for a Change

Start by talking with your doctor. He or she may be able to determine if the way you feel is related to poor diabetes control or another factor. If your diabetes isn't well managed, your blood glucose may drop during the day, so your anxiety and fatigue may have a medical cause. These same low levels may cause you to eat too much. At night, low blood glucose can disturb your sleep. Working together, you and your doctor may be able to ease the stress, denial and anger with a customized treatment regimen.

Support groups also can be a tremendous source of help. You'll meet other people who have already faced the same issues. It's likely they will share ideas that helped them. Before adopting any idea that alters your treatment plan, be sure to talk with your doctor. Others can serve as the role models you need to get on track with your diabetes self-management.

Sometimes one-on-one counselling can make a difference. Whether you see a social worker, psychologist, counsellor or psychiatrist, these professionals can help you find new and healthier ways of living a full, rewarding life, despite diabetes.

Symptoms of Depression

Depression, whether it's the result of feeling a loss of control or directly the result of your diabetes, has specific symptoms. These include:

- loss of energy
- inability to enjoy life
- increased appetite or weight loss
- trouble concentrating
- difficulty sleeping, or changes in sleeping patterns
- memory lapse
- trouble getting things done
- moodiness, irritability or restlessness
- feelings of emptiness or worthlessness
- suicidal thoughts.

Questionnaires are good tools to use in finding the source of problems like depression. The one that follows was developed by myself and David MacDonald, DO, a colleague who shares my interest in getting to the root of problems.

Solutions, not Problems with the 11—11 Scale

This scale focuses on 11 layers of interaction that exist between yourself and others. At any time you can take your 'psychological temperature' or self-reading of where you are on the 11—11 scale.

Although 11 points on all 11 scales is the ultimate target, 121 points in all, most of us will never reach this level of satisfaction. Using this scale gives you an objective number to tell you how you are feeling at any point in time.

> In the test that follows, answer spontaneously without thinking too much, since this should be a measure of your 'gut feeling'.
>
> An answer of '1' indicates that the way you feel about the question is poor, low, bad, couldn't be worse. An answer of '11' means that the way you feel about the question is great, excellent, couldn't be better. Try it for yourself and see how close to 121 you come:
>
> | Questions | Answers (1—11) |
> |---|---|
> | How do you feel about you, yourself? | ☐ |
> | How do you feel about your soulmate? | ☐ |
> | How do you feel about your family? | ☐ |
> | How do you feel about your friends? | ☐ |
> | How do you feel about your work? | ☐ |
> | How do you feel about your environment? | ☐ |
> | How do you feel about your spirituality? | ☐ |
> | How do you feel about your nutrition? | ☐ |
> | How do you feel about your level of physical activity? | ☐ |
> | How do you feel about your level of control? | ☐ |
> | How do you feel about your level of hope? | ☐ |

Interpretation

11—30   You are probably experiencing conflict in many areas of your life, which can interfere with your overall health and wellness. If you score at this level over several days, you may consider seeking help from a professional trained in lifestyle management.

31—60   You probably have areas which are acceptable but many which are less than desirable. Look at specific low scores to see if there are any ways to bring the scores up. Perhaps your sense of control could increase by increasing the amount of exercise you do or improving your nutritional choices.

61—90   Your scores are more positive than negative, which means you are on the road to improved health and wellness. Keep trying to improve by identifying the low points and bringing them up.

91—121   If you fall into this range, keep up the good work. You probably have a good sense of what it takes to identify conflict and resolve it appropriately.

This scale not only gives you a total stress score, it also helps you isolate stressors in your life. If, for instance, you score yourself low on nutrition, then better eating would lower stress in your life. Or if you rate yourself low on your feelings about your family and also have a low sense of control, then family counselling might be the answer to reducing stress in your life.

The Language of Denial

Denial can be the greatest enemy to your long-term good health. Pay attention when you hear yourself say:

- I feel fine. I know when my blood sugar is out of control.
- I can't afford the time to see a doctor, dietitian, etc.
- It doesn't matter if I miss testing my blood glucose today.

- That food looks good, a little more won't hurt me.
- It's too much trouble to test my blood sugar at work.
- That sore on my foot will heal.
- I can't be bothered cooking separate meals for my family and me.
- I smoke or I drink alcohol. It's who I am and I'm not giving it up.

Finding and pursuing fulfilling activities is important for everyone, but especially if you have diabetes. Here are some ideas about how to turn your attitude and life onto a more positive path:

- Make new friends and nurture old friendships. People who interact in a positive way with others tend to live longer, happier lives.
- Foster family ties and traditions. Seek ways to build continuity and support in your life through family and family traditions.
- Get involved. Join a community group, volunteer, help out and reach out. The more you think of others, the less likely you are to feel sorry for yourself.
- Get busy. Pursue a hobby you enjoy and which allows you to interact with others. Sign up for a class. Learn or teach someone else a new skill.
- Get fit. Talk with your doctor about appropriate fitness activities.

Stress at Work

Your work is a part of who you are and so is your diabetes. One Swedish study in the *Journal of Occupational Medicine* found the amount of support from supervisors and co-workers, and the ability to control the pace, volume and type of work all play into the amount of stress you feel at work. The same study found a connection between repetitive and monotonous work and workers' difficulty in relaxing, even after the work day ended.

Exactly how much your diabetes affects your work life is very individual. Some factors include:

- your hours. Do you work shift work or 9—5?
- your work style. Does your work easily accommodate snacks, breaks and checking your glucose levels?
- your work environment. Are the people you work with a supportive team, or is it each person for him- or herself?
- your profession. Does your work allow you some control over the amount of work you have?
- your benefits. Does your job have such benefits that make managing your diabetes well possible?
- job satisfaction. Do you love your job or profession or is it just a means to a pay packet? Is it rewarding and satisfying? Do the pay and benefits meet your needs?

Depending upon how well controlled your diabetes is, the disease may or may not be an issue at work. Talk with your diabetes management team about your employment. Keep careful records about your blood glucose levels and work activities. If appropriate, talk with your boss about changes at work you may need to make to maintain your job performance while managing your diabetes.

Constructive Ways to Cope with Stress

- exercise
- relaxation
- hobbies
- visualization
- therapy
- spirituality or faith
- support groups

Destructive Ways to Cope with Stress

- missing meals or overeating
- using too much alcohol, prescription medication
- using illicit drugs
- missing appointments with your doctor
- holding your feelings in
- isolating yourself
- abusing family and friends
- not exercising

Stress on the Road

Travel, whether for business or pleasure, can present specific challenges for people with diabetes. The excitement of preparation and planning, of visiting new places, meeting new people and eating out can cause your blood glucose to rise. Likewise, changing time zones and schedules can make diabetes management difficult. Talk with your doctor about any special considerations you may have, and use these tips to help you feel the best while on the road.

- Keep your diabetes supplies with you at all times. Never check them with baggage. Pack twice as much as you normally need in case of delays or an emergency.
- Carry snacks and glucose tablets (and when visiting certain countries, purified water) with you at all times. This is especially important if your travel schedule is a rigorous one.
- Talk with your doctor about adjusting your food plan to accommodate time zone shifts and regional differences in food.
- Wear your diabetes identification at all times.
- Take at least two pairs of shoes that are comfortable and well broken in. Change your shoes and socks at least once a day and take extra care in checking your feet for abrasions, blisters and chafing.

- If travelling in countries where English isn't commonly spoken, learning how to say a few basic phrases can be a life-saver, such as 'I have diabetes.' 'Please get me a doctor.' 'May I please have sugar or a glass of orange juice?'

The International Association for Medical Assistance to Travellers (IAMAT) has set up centres in 25 countries staffed with English- or French-speaking doctors who are on 24-hour call. IAMAT also publishes a directory of medical centres and associated doctors who have agreed to a set payment schedule for IAMAT cardholders. The organization can also provide information about food, climate and sanitary conditions in countries you plan to visit. For more information please contact: IAMAT, 57 Voirets, 1212 Grand-Lancy-Geneva, Switzerland or 40 Regal Road, Guelph, Ontario, Canada N1K 1B5 (fax: 519 836 3412; e-mail: iamat@sentex.net).

CHAPTER EIGHT

An Ounce of Prevention

The Programme

What? Live with the confidence of a person who knows he or she will not get type 2 diabetes.

How? By following a 5-step prevention programme:

Step 1 Know your risk of type 2 diabetes.
Step 2 Watch your weight.
Step 3 Eat from a healthy list of body-friendly foods.
Step 4 Exercise on a daily basis with activity you enjoy.
Step 5 Take HCA as a prevention against type 2 diabetes.

When? Think about these steps to prevention every day to make changes easier.

Why? Because type 2 diabetes can be prevented, controlled and *cured* using these five simple steps.

Prevention is the best medicine, especially when it comes to type 2 diabetes. Researchers and medical professionals have identified the following as essential to preventing or lowering your risk of developing diabetes:

Know your risks for developing diabetes. Have your blood glucose level checked regularly. After age 45, have it checked every three years; more often if the resulting reading is high or if you have a family history of diabetes.

Those at higher risk – Afro-Caribbeans and those of Asian or Hispanic descent – should be tested earlier and more often.

Keep your weight in a range normal for your height and age. Eat a well-balanced diet that is low in calories, fat and refined carbohydrates.

With your doctor's approval, develop an exercise programme including 20 to 30 minutes of aerobic activity daily, plus stretching and strengthening activities.

Take HCA as a means of improving your insulin receptivity. This will help you lose weight, and it will help you feel more energetic, which will make exercise easier and more desirable to do.

FIGURE 8.1

A Chart Review of 353 Patients in a Large California HMO Revealed There Is Serious Lack of Preventive Care in Diabetes Patients

- 56% of patients had no HbA1c.
- 39% who had HbA1c were over 10% (i.e., blood glucose greater than 250)
- 65% had no fasting blood sugar (FBS).
- 76% had no lipid measurement.
- 52% had no urinary protein measurement.
- 78% had no ophthalmology referral.
- 94% had no foot exam.
- 45% had an emergency department visit for diabetes.

It Is Important to Keep Checking the Following Regularly:

- Haemoglobin A1-c (HbA1c) every 3 months
- Fructosamine (glycoprotein) testing every 3 to 4 weeks.
- Assure that blood glucose is less than 250.
- Fasting blood sugars.
- Measurement of serum lipids.

- Ophthalmological referral for retina examination yearly.
- Careful foot exam – podiatrist recommended yearly.
- Prevention and education to avoid ER visits unless necessary.

Source: A. L. Peters, 'Quality of Outpatient Care to Diabetic Patients: A Health Maintenance Organization Experience.' *Diabetes Care* 19 (6), 1996; 601–606.

Experts estimate that almost all type 2 diabetes could be avoided, controlled or *cured* by following these guidelines. Although more and more people are becoming sedentary and overweight, surveys have shown fewer risk factors for diseases like type 2 diabetes in populations educated about safe lifestyles. By reading this book you know how to make yourself healthier and are more likely to do it. Congratulations! By following these steps you will lead a happier and healthier life. You will also help friends and family make changes that will improve their lives, too.

Step 1: Know Your Risk of Type 2 Diabetes

About 2 per cent of the population has diabetes. Every year some 50,000 new cases are diagnosed, with 90 to 95 per cent of these among people over 20 with type 2 diabetes. Half of these are unaware they have diabetes. That's because it's possible to have mild symptoms for years before finally seeking medical help. Symptoms of type 2 diabetes include:

- unusual thirst
- frequent desire to urinate
- blurred vision
- frequent, prolonged fatigue with no apparent cause
- leg pain
- unexplained weight loss.

Most people with type 2 diabetes are insulin-resistant. This means that although their pancreas is producing insulin, either their cells are unable to obtain it or insufficient insulin is being produced. No matter the cause, without insulin's effect to meet your body's needs, glucose levels rise and diabetes and its many complications can result.

The Gene Connection

The link between a family history of diabetes and your risk of developing it is clear, but not necessarily inevitable. The link is certainly stronger for type 2 than type 1. Researchers have yet to identify and isolate a single type 2 diabetes gene. What they do know is that if you have type 2 diabetes and are an identical twin, there is a 60 to 75 per cent chance your twin will also develop the disease.

The high incidence of diabetes among certain minority groups also supports the genetic connection. Those of Afro-Caribbean and Asian descent get diabetes more often compared to Caucasians. It is also interesting to note that Native Americans have the highest rate of diabetes in the world, and Mexican Americans, who share genes with Native Americans, have higher rates than Cuban Americans, among whom less inter-cultural breeding has occurred.

Some researchers believe the same genes that cause obesity may play a role in diabetes. Scientists compared the Arizona Pima Indians with the Mexican Pima Indians, two tribes related by genetics but separated by culture and national borders. The Arizona Pimas live 30 miles south of Phoenix on a reservation with electricity, running water, televisions and cars. Many are unemployed, and most of those who do work have sedentary jobs. Their diet, like that of many Americans, is heavy on fatty, processed foods.

The Mexican Pimas, on the other hand, live in the Sierra Madre mountains of Mexico. They have no electricity or running water. They have no modern conveniences, and all jobs require heavy physical labour. Their traditional diet is very low in fat and incorporates lots of beans and hand-made tortillas.

When scientists compared the two groups' diabetes risks, the difference was startling. Scientists found that while Mexican Pimas and their non-Pima neighbours in Mexico had similar rates of diabetes – 6.4 per cent vs 3.4 per cent – the rate for Arizona Pimas was 38.2 per cent. Scientists peg the increased diabetes risk for the Arizona Pimas to obesity.

The University of Washington developed this analysis to help you calculate your risk of developing diabetes. Test yourself.

I had a baby weighing more than nine pounds at birth
  or had diabetes during my pregnancy.   Yes: 6. No: 0.
I have a parent(s), sister or brother with
  diabetes.   Yes: 3. No: 0.
I am of Afro-Caribbean or Asian descent.   Yes: 3. No: 0.
I am overweight.   Yes: 3. No: 0.
I have been told I have a high blood
  sugar level.   Yes: 6. No: 0.
I am between 45 and 64.   Yes: 1. No: 0.
I am under 65 and I get little or no exercise
  during the usual day.   Yes: 3. No: 0.
I am 65 or older.   Yes: 3. No: 0.
Total your score and check your risk.

1—5   You probably are currently at a low risk of developing
      diabetes.
6 or more   You are at high risk of having or getting diabetes.

Other Causes for Concern

There are other causes of type 2 diabetes not covered in this test. Recent research has linked chemical exposure and the use of certain medications to type 2 diabetes and other forms of diabetes as well. Exactly why this happens isn't known, but medical science is working hard on that question as you read this book.

Several studies have pointed to a connection between dioxin, a deadly chemical once used worldwide in pesticides and in the defoliant, Agent Orange, and diabetes. One study compared the occurrence of diabetes among Vietnam veterans who were exposed to Agent Orange during duty in Vietnam with those who had not been exposed. The exposed veterans were 50 per cent more likely to develop diabetes.

Another study points to a possible link between dioxin and the startling jump in type 2 diabetes among children around the world. Researchers speculate a connection between obesity, diabetes and the body's storage of chemicals such as dioxin in the fat cells.

People who take protease inhibitors, a powerful drug used to treat people with AIDS, appear to have an increased risk of diabetes. The actual risk of diabetes from protease inhibitors is not known because many of the estimated 150,000 HIV-positive Americans on the drug may have undetected hyperglycaemia. The US Food and Drug Administration speculates the risk is between 1 in 100 and 1 in 1,000.

Contraceptives and Pregnancy

The jury is still out on whether the use of oral contraceptives contributes to the development of diabetes. One study at the University of Southern California Medical School found that those who have had gestational diabetes or had a family history of diabetes and take certain birth control pills after delivery may treble their risk of developing diabetes later in life.

The women at greatest risk for developing type 2 diabetes years later used the suspect contraceptives for six months to a year after deliveries complicated by gestational diabetes, according to a study which focused on almost 1,000 obese Hispanic women who recovered from gestational diabetes four to six weeks after delivery. Over the next seven years, 56 per cent of the women used some form of hormonal contraception. The contraceptives identified as producing the greatest risk of type 2 diabetes were progestin-only pills, progestin implants and oestrogen-progestin combinations with the strongest progestins.

A different study from Harvard Medical School placed the blame squarely on obesity. Women who have diabetes or who are at risk for developing it – either through a previous gestational diabetic episode or a history of diabetes in their families – should talk with their doctors about taking combination contraceptives with the lowest dose and potency of progestin. Of the contraceptives studied, the combination pill containing norethindrone (0.4 mg) appeared to be the safest and did not appear to increase risk.

Step 2: Watch Your Weight

The number of people with type 2 diabetes has been increasing dramatically. It has jumped almost 50 per cent since 1983 and trebled since 1958.

The International Diabetes Institute predicts the number of people with diabetes worldwide is expected to double by the year 2010. Diabetes has been called epidemic in many regions of the world, especially the west. Currently 123 million people worldwide have diabetes. By 2010, the figure is expected to grow to 220 million. Experts point to increasing weight and sedentary lifestyles as the catalysts.

About nine out of ten people with type 2 diabetes are overweight. According to the World Health Organization, by the year 2025 more than 300 million adults worldwide are expected to have diabetes. Insulin resistance seems to be the cause. Insulin resistance – the body's failure to respond to insulin – develops as weight increases. Experts believe that at least 80 per cent of type 2 diabetes could be eliminated by getting rid of obesity.

Losing as little as 10 pounds and keeping it off can cut your risk of developing diabetes. Modest weight loss works even for those who would otherwise be at a high risk of developing the disease, a study from the University of Pittsburgh has shown. Their study focused on 157 overweight men and women between the ages of 39 and 55, all of whom had at least one parent with type 2 diabetes, though none showed signs of having diabetes themselves.

After six months, scientists found every participant who made lifestyle changes and lost weight had a reduction in average blood glucose levels. In fact, researchers calculated that a 10-pound weight loss maintained for two years could reduce the risk of developing diabetes by 31 per cent. Where you carry weight also seems to be a factor in developing diabetes. Extra weight above the hips is riskier than fat in the hips and thighs.

As everyone knows, losing weight is moderately easy. Keeping it off is tough. Yo-yo dieting – losing and regaining weight – appears to be another factor increasing the risk of developing type 2 diabetes. In a study published in *Obesity Research*, rats on a yo-yo diet that alternated adequate caloric intake with severe caloric restrictions developed insulin resistance as they aged, a precursor of type 2 diabetes. Instead of yo-yo dieting, the answer is to make slow, permanent weight loss your goal via a low-fat diet and regular exercise.

Not just adults are at risk. Children as young as eight are developing type 2 diabetes and at an increasing rate. Researchers at the Arkansas Children's Hospital tracked 50 children with type 2 diabetes and compared them with 50 children with type 1 diabetes.

Type 2 children, researchers found, were nearly all obese; more than 30 per cent already had high blood pressure. Of the children with type 2 diabetes, 74 per cent were black and 24 per cent white. In contrast, 18 per cent of the type 1 children were black and 82 per cent white, a statistical make-up not unlike the demographics of Arkansas' population.

The finding startled researchers, because type 2 diabetes usually develops in people over 30, and previously was considered rare in children. Type 2 diabetes is associated typically with obesity, and the latest figures show obesity in children and teens has increased dramatically in recent years.

This is not the first study to uncover rising rates of type 2 diabetes in children. Two studies in San Diego and one in

Cincinnati came to similar conclusions about children in those areas, suggesting this is widespread phenomenon; some experts speculate it may even be worldwide. At a recent meeting of the International Diabetes Federation Congress in Helsinki, Finland, experts agreed that the world faces a global explosion in diabetes unless people who are obese can lose weight.

Body Weight for Adults

Keeping your weight in a range normal for your height and age is the single most important thing you can do to reduce your risk of diabetes. The following are recommendations from the US Department of Agriculture.

Height in inches, without shoes	Weight in stones without clothing	
	Age 19—34	Age 35 and Over
5'0	7—9	7½—10
5'1	7 st 3—9 st 6	8—10 st 3
5'2	7½—9 st 11	8 st 3—10 st 8
5'3	7 st 9—10 st 1	8½—10 st 12
5'4	7 st 13—10 st 4	8 st 10—11 st 3
5'5	8 st 2—10 st 10	9—11 st 8
5'6	8 st 6—11 st 1	9 st 4—11 st 13
5'7	8 st 9—11 st 6	9 st 8—12 st 4
5'8	8 st 13—11 st 10	9 st 12—12 st 10
5'9	9 st 3—12 st 1	10 st 2—13 st 1
5'10	9 st 6—12 st 6	10 st 6—13 st 6
5'11	9 st 10—12 st 11	10 st 11—13 st 12
6'0	10 st—13 st 2	11 st 1—14 st 3
6'1	10 st 4—13½	11 st 5—14 st 9
6'2	10 st 8—13 st 13	11 st 10—15
6'3	10 st 12—14 st 4	12—15 st 6
6'4	11 st 2—14 st 9	12 st 5—15 st 12
6'5	11 st 6—15 st 1	12 st 9—16 st 5
6'6	11 st 10—15 st 6	13—16 st 10

Step 3: Eat from a Healthy List of Body-friendly Foods

The right foods can make your body produce the right amount of insulin.

Mounting evidence suggests diets low in fibre and high in the 'wrong' carbohydrates – refined items such as white bread and granulated sugar – more than double the risk of developing diabetes.

According to a large prospective study, different foods trigger different amounts of insulin secretion, even if the foods contain the same amounts of carbohydrate. The difference (glycaemic index) lies in how easily the starch is digested – for example, white bread and mashed potatoes are very digestible and lead to substantial insulin secretion; dark bread, yoghurt and fruits are less digestible and trigger less insulin. Eating too many of the wrong carbohydrates too often can lead to elevated insulin levels, a risk factor for diabetes. To lower your risk of developing diabetes, the less refined the food you eat, the better, researchers say.

Everyone knows fibre is good for you. Lack of dietary fibre has long been considered a risk for several chronic diseases, including diabetes. Researchers fortified this link by studying more than 65,000 women over six years. Taking into account obesity, sedentary lifestyles, genetics and other risks, the results show fibre intake, especially cereal fibre, reduces the incidence of diabetes. Fruits, vegetables and grains, especially unrefined cereals, are excellent sources of 'good' fibre, unrefined carbohydrate and some magnesium.

A study from the Netherlands found that men who ate healthy meals, as well as exercised, were less likely to develop insulin resistance. Researchers working with 389 elderly men who did not have diabetes found insulin levels were highest among those who got the least exercise and whose diets included the largest proportion of saturated fats and alcohol. High insulin levels are a sign of insulin resistance. Insulin levels decreased as the amount of high-fibre carbohydrates increased. Insulin levels also declined in men who consumed fats as polyunsaturated oils.

Another study, this one from Stanford University, found that subjects who took three to five times the recommended daily average (RDA) of vitamin A were 40 per cent better able to move glucose from their blood to their cells.

Vitamin A is a fat-soluble vitamin commonly found in liver, fish liver oils, whole and fortified milk and eggs as well as orange fruits and vegetables and green leafy vegetables. It's essential for vision, a healthy immune system and bone building. Because of the health risks associated with overdosing on vitamin A supplements, it's a smarter strategy to maximize your intake of vitamin A-rich foods.

Step 4: Exercise on a Daily Basis with Activity You Enjoy

Regular exercise can make your muscles use glucose correctly.

Exercise – almost any kind – can reduce your risk of developing diabetes. So whether you're a dedicated lap swimmer or walker, a serious gardener or a frequent stairs-instead-of-the-lift person, getting active and staying active pay off.

Don't be confused. Being physically active isn't the same thing as fitness. According to researchers at Johns Hopkins University, it isn't necessary to devote your life to physical fitness in order to gain health benefits. Their research has shown a reduced disease risk comes with just modest amounts and intensity of exercise – as little as 30 minutes of activity per day.

Here's an example of how that 30 minutes can pay off. The Honolulu Heart Program studied 6,815 men, looking for a link between exercise and diabetes. The men, who were diabetes-free when the study began, provided detailed information about their lifestyle, including their exercise programmes and day-to-day activities such as walking or gardening.

Six years later, researchers found diabetes developed most frequently in those who were the least active. Compared with the active men, the inactive men's risk of developing the disease was 51 per cent greater.

For older adults, who make up a substantial proportion of the people with diabetes, regular, doctor-approved activity may be even more beneficial than dietary changes in reducing the risk of developing diabetes, according to the University of Washington. An estimated 18.4 per cent of people 65 and older have diabetes.

The good news about increasing your activity level is that many people see quick improvement of insulin resistance. A University of Michigan study found that one hour a day of walking or stationary bicycling reduced the insulin resistance of seven of eleven African-American women.

Take note – 'activity' is different from the single-minded pursuit of strong biceps, four-minute miles or rock climbing. To improve your health, your goal should be to incorporate 30 minutes of accumulated activity every day. The payoff is broader than just lowering your risk of developing diabetes. It also includes the potential to lower your blood pressure and reduce your risk of cardiovascular disease and osteoporosis, not to mention helping you sleep better, think more clearly and feel less stressed.

With regular activity comes an increase in muscle mass. The greater the percentage of muscle, the more effectively you burn calories in food and the more sensitive your body is to insulin.

Step 5: Take HCA as a Prevention Against Type 2 Diabetes

In theory, type 2 diabetes is easy to prevent because it is largely a disease of lifestyle. That means simply that this disease can be caused by the habits you have developed in your life, especially ones that make you eat too much and exercise too little. By following the Diabetes Cure programme *even if you haven't got diabetes*, you are changing these habits and preventing diabetes and many other diseases of lifestyle. By following the simple plan outlined in this chapter, for example, you can prevent such afflictions as heart disease, strokes, arthritis, gout and possibly even some forms of cancer. The US National Institutes of Health says that six out of ten diseases are caused by lifestyle and would never happen if diet, exercise and other lifestyle guidelines are followed.

Taking HCA in the amounts recommended in this book is harmless and will help you lose weight, which is the number one cause of type 2 diabetes. If you weigh too much for your height, as indicated by the height/weight chart on page 167, consider following this programme as a preventative against diabetes and as a means of improving overall health.

'Unteaching' Habits

If so many diseases are so easy to cure, why are afflictions like type 2 diabetes on the rise? The answer lies in habit control. Our habits have been with us our entire lives – some even longer when you consider that they were handed down to us by our parents. So the children of vegetarians usually grow up to become vegetarians, just as meat-eating families spawn meat-eating children.

It is difficult to change the habits with which we were raised because they are what we know and are part of our identity. Many people, especially those who are overweight, feel guilty about their habits. They, more than anyone, would like to change their habits and lose weight, but it requires a constant vigilance that they sometimes cannot maintain.

Frankly, it is difficult enough to make lifestyle changes without feeling guilty about failing. I have had patients cry in my treatment rooms because they have failed to maintain an exercise programme or avoid their favourite fattening foods. I'll say to you what I say to them:

Our habits have been taught to us by many people throughout the years. We learned to eat too much ice cream from someone just as we learned to sit in front of the television from someone. Lifestyle change and habit control is a long, slow process that takes time. Ask your family and friends for help and ask them to be understanding.

As far as I am concerned, slow and realistic change is the only way to make lifelong lifestyle changes.

CHAPTER NINE

Diabetes Disease Review

Diabetes affects more than your blood glucose levels. It's a chronic condition which, when left unchecked or poorly managed, can devastate virtually every organ and system of the body.

It can be difficult to understand and accept that what you eat today, the amount of exercise you get this week as well as how you manage your diabetes with a programme like the one outlined in this book can make a difference in your health. But it's true. Every day of diligent blood glucose control, of a balanced diet plus exercise is an investment in reducing the risk of complications and improving your potential for long-term quality of life.

For many years, medical professionals were able to see the benefits of diligent diabetes control in their patients. It wasn't until the close of the US Diabetes Control and Complications Trial in 1993, however, that research and study supported and confirmed this intuition.

Study participants who maintained rigorous control of their glucose cut their complication risk in half. They had 76 per cent less eye disease, 60 per cent less nerve damage and 35 to 56 per cent less kidney damage. Researchers believe people with type 2 diabetes can reap similar benefits from tight blood glucose control coupled with lifestyle changes such as weight loss and exercise. The benefits of tight control were so obvious that the study was ended early and the findings announced so that others could benefit from them, too.

Changing for the Better

Even if you are already experiencing some diabetes complications, take heart and take action. Strict blood glucose control and smart lifestyle choices can minimize complications and may prevent the development of additional ones. Those who feel that tight control just doesn't work for them can still reduce the risk of developing life-threatening complications by keeping their weight under control, exercising and managing their disease.

The fact is, nearly every complication of diabetes comes from having too much glucose in your blood. A good metaphor is what happens when you put lots of sugar in a glass of tea or cup of coffee. The beverage becomes sticky and thick. That same dynamic is at work in most diabetes complications.

The thickness causes arteries and veins to narrow and blood flow to be reduced. It also can damage nerve cells and interrupts or stops such electrical messages, such as sensations of pain or heat, that your cells send to all parts of the body. Without sufficient or smooth system-wide blood flow or in the absence of efficient nerve conductivity and performance, a wide range of complications, from atherosclerosis and heart attack to blindness, impotence and nerve damage can occur.

Complications Overview

The complications of diabetes are diverse and widespread. Some are common and well known, and many are related. Others occur on a limited basis, often in the presence of pre-existing complications. Here's an overview of the most frequently experienced diabetes complications, how you can prevent them and some of the common methods of treatment.

Hypoglycaemia

The possibility of hypoglycemia is a part of having diabetes, especially if you use insulin or other diabetes medications. It occurs most often before meals, when insulin is peaking, or during or after vigorous exercise. You can even experience hypoglycemia when you're asleep. It can be especially problematic for pregnant women.

When blood glucose levels drop too low, you can feel dizzy and shaky. If untreated you can lapse into unconsciousness, have seizures and go into a coma. This is a true medical emergency that can be life-threatening if left untreated.

Hypoglycaemia Symptoms

weakness or fatigue	rapid heartbeat
shaking and sweating	dizziness and impaired vision
hunger	headache
loss of coordination	nausea
nightmares	irritability, anger and confusion.

In a severe occurrence, it can leave you so disorientated or belligerent that you can't help yourself or even ask for help from others.

These symptoms may be misinterpreted as stemming from drunkenness, drug abuse or a psychiatric disorder. That's why it's important to work to avoid hypoglycaemia as well as to wear diabetes identification, so that your condition can be spotted and treated even if you can't speak for yourself.

People who take blood pressure medications known as ACE inhibitors should be particularly alert for hypoglycaemia. A researcher at the Ninewells Hospital in Dundee, Scotland, analysed hospital admissions for hypoglycaemia and ACE inhibitors, a type of blood pressure medication. He found that people using the blood pressure medications were three times more likely to be hospitalized with severe hypoglycaemia than those who did not take the drugs.

If you use ACE inhibitors to manage your high blood pressure, talk with your doctor about how to limit your hypoglycaemia risk. Never stop taking your high blood pressure medication without first talking with your doctor as hypertension also can be a complication of diabetes which contributes to other diabetes-related conditions.

TREATMENT

Because your body's use of insulin is determined by many factors such as stress, exercise, the kinds of food you eat, when you eat it and how much, and the overall state of your health, monitor your blood glucose levels carefully and frequently. Be alert for symptoms which can vary, and take action immediately by eating or drinking a sugar that can enter your digestion and bloodstream quickly. Possibilities include glucose pills or gels. Don't use sweets or chocolate, as the fat in these can slow sugar absorption.

Retest your blood glucose levels about 15 minutes later. You may need to add another small dose of sugar, but take care to avoid a too-rapid rise in blood glucose which can occur from too much sugar.

After repeated bouts of hypoglycaemia, some people lose the ability to detect when one is coming on; others have difficulty detecting it even with tight control. Protect yourself by eliciting the help of family, friends and co-workers. Educate them about the symptoms, including the possibility of anger and out-of-control behaviour, and the appropriate actions to take. Make clear that hypoglycaemia is a life-threatening condition and that if they are unable to intervene, the next step is to call for emergency medical assistance. Your life could be in their hands.

Hypoglycaemia at night is a problem which can be particularly perplexing for people with diabetes. Traditionally, treatment has included having a snack before bed to prevent low glucose levels during the night. At Washington University School of Medicine, researchers studied 15 people with type 1 diabetes and found that when taken at bed time, alanine, an amino acid, stimulated release of glucagon, while terbutaline normalized blood glucose. Thus, these medications may be a way to avoid hypoglycaemia at night.

Hyperglycaemia

Hyperglycaemic hyperosmolar nonketotic syndrome (HHNS) is most common in people with type 2 diabetes, especially those who do not use insulin. Symptoms include:

extreme thirst and a parched mouth	confusion or sleepiness
dry skin; inability to perspire	blurred vision
hunger	nausea
frequent urination	extremely high blood glucose levels

HHNS occurs when blood glucose levels continually increase. Urine output speeds up; as a result, you become dehydrated. Over time – sometimes days or even weeks – you may become confused and unable to ask for water or make it to the bathroom. As the blood becomes thicker due to the glucose build-up and dehydration, you may have seizures, go into a coma and even die.

TREATMENT

Drink plenty of alcohol-free liquids throughout the day. Checking your blood glucose levels once a day can alert you to impending HHNS. If your level is greater than 350 mg/dl, call your doctor; greater than 500 mg/dl, go immediately to hospital. People who also take medications such as steroids, diuretics, Tagamet or beta blockers may need to test their blood glucose more often.

HHNS is especially a risk for people who receive intravenous feedings or peritoneal dialysis. About one-third of HHNS occurs in people in nursing homes, who become dehydrated because they must wait for staff to provide liquids. Frequent blood glucose monitoring is the best solution.

FIGURE 9.1

An Aspirin a Day Keeps the Doctor Away

Diabetes carries a two to four times higher risk for early death from heart disease.

The ADA recommends that 81 mg to 325 mg of enteric-coated aspirin be used:

Secondary Prevention:
   Patients with large vessel disease

Primary Prevention:
Patients with the following risks:
   High blood pressure
   Obesity
   High urine protein
   Cigarette smokers
   Total cholesterol greater than 200 mg/dL
   LDL cholesterol greater than 130 mg/dL
   HDL cholesterol less than 40 mg/dL
   Triglycerides greater than 250 mg/dL

Cardiovascular Complications

If you have diabetes, you're two to four times more likely to develop cardiovascular disease and five times more likely to have a stroke than those without diabetes. More than half the deaths in people with diabetes are due to heart disease.

It all comes down to blood flow. Blood transports oxygen, glucose, nutrients and other essential substances to the tissues of the body. If that flow is interrupted or slowed due to uncontrolled blood glucose levels, tissues and vital organs suffer and even may die.

Atheroscolerosis

Unchecked blood glucose levels change the chemistry of the bloodstream. Those changes can lead to the clogging of vessels. This clogging may be the result of too many lipids – cholesterol and triglycerides – in the blood.

This build-up can cause these substances to cling to the sides of veins and arteries, narrowing them. Add to that the stickiness of too much glucose in the blood and the process is accelerated. Platelet changes in the blood due to uncontrolled blood glucose levels also can contribute to vessel narrowing.

TREATMENT

A low-fat, high-fibre diet that minimizes the fat, cholesterol and triglycerides in the blood plus exercise can make a difference. Medications can help control the build-up of substances that clog the circulatory system.

Treatment with large doses of vitamin E, a potent antioxidant, also may offer help. In a study of 28 people with diabetes at the Center for Human Nutrition at the University of Texas Southwest Medical Center, researchers found 1,200 IU of vitamin E contributed to a decreased risk in heart attack. At the same time, it also reduced rates of LDL or so-called 'bad' cholesterol. The study builds on the findings of other research involving people without diabetes, which appeared to show that vitamin E reduces LDL levels by blocking its oxidation.

Before increasing your vitamin E intake from the recommended daily average of 10 mg, talk with your doctor to be sure intensive vitamin E supplementation is right for you.

Angina

More commonly known as chest pain, angina occurs when blocked vessels restrict blood flow to and from the heart.

TREATMENT

Medications that widen vessels and arteries or reduce the build-up of plaque in the bloodstream.

Heart Attack

Occurring when blocked vessels to the heart are completely blocked by deposits, heart attack is a significant issue for people with diabetes.

TREATMENT

Medications that widen vessels and arteries or reduce the build-up of plaque in the bloodstream are the most frequent solutions. In advanced cases, arterial bypass surgery to create new paths for blood flow may be necessary. The bypass, itself, is created by removing vessels from another part of the body, usually the legs, and using them to create new routes for blood flow to and from the heart.

People with diabetes heal more slowly than those without the disease and have higher rates of infection, both of which can be serious post-surgical complications for people with diabetes.

High Blood Pressure

Nearly 60 per cent of people with type 2 diabetes have high blood pressure. Hypertension means your heart has to work harder to move blood to and from the heart; that extra work can damage the circulatory system. Eventually, due to the build-up of atheroma – fatty tissue – veins and arteries may be completely blocked.

TREATMENT

Maintain your weight in a normal range and eat a low-fat, low-sodium diet; with your doctor's approval, exercise 20 to 30 minutes, three to four times a week or more. Above all, stop smoking. All of these can help lower blood pressure.

Medications are also available to lower blood pressure itself, as well as reduce blood clotting and lower cholesterol. Surgical interventions may include balloon angioplasty to open blocked vessels, and arthrectomy to bore away blockages.

Elevated Cholesterol

Diabetes can change the number and make-up of proteins that deliver lipids (fats) to the cells. Elevated cholesterol and blood lipids can also accumulate if you eat too much fat. With too much blood glucose in the bloodstream lipoproteins become extra sticky.

TREATMENT

Rigorous blood glucose management and a diet that's low in fat and high in fibre, plus regular, doctor-approved exercise can go a long way towards good cholesterol management. In addition, researchers have found Simvastatin, one of the new classes of cholesterol-lowering drugs, can significantly lower cholesterol levels and dramatically reduce heart attack risk.

In one European study, 202 people with diabetes were part of a 4,444-person study of the drug. Results indicated that Simvastatin appeared to reduce LDLs, increase HDLs and lower cholesterol; people with diabetes benefited the most.

Simvastatin lowered their risk of heart attack by 55 per cent, compared with 32 per cent among persons without diabetes. Additionally, the death rate from heart attack among participants with diabetes dropped to 43 per cent, compared to 29 per cent among people without diabetes. Researchers believe Simvastatin also has no effect on blood glucose levels, making it particularly appropriate for people with diabetes.

Stroke

Rates of stroke are two to four times higher among people with diabetes than in the general population. Strokes commonly occur when a small chunk of the plaque that has built up on the lining of vessels and arteries breaks free and circulates in the body until it reaches a passageway through which it cannot pass. This blockage can result in a stroke.

TREATMENT

Keeping blood pressure under control, eating a low-fat, high-fibre diet and exercising regularly with your doctor's approval are the best ways to reduce stroke risk. Your doctor also may recommend medications to lower blood pressure and dissolve clots. How well you relate to stressful situations also has an effect on stroke risk.

Intermittent Claudation (IC)

More commonly referred to as leg pain, IC occurs when vessels and arteries in the legs are partially or fully blocked or restricted.

TREATMENT

Exercise and stopping smoking are the best preventions. Your doctor may also recommend medications commonly used to treat blocked vessels and arteries as well as surgery to remove blockages.

Dementia

The spectre of developing dementia later in life as a result of uncontrolled blood pressure is a complication which has recently gained attention.

TREATMENT

Keeping blood pressure at or near normal levels through medication, exercise and diet may help.

Vision

Diabetes is the leading cause of new blindness among adults. Indeed, people with diabetes are four times more likely to become blind than people without the disease.

Diabetic Retinopathy

This is the most common cause of blindness among people with diabetes. Caused by damage to the vessels supplying blood to the retina, most people don't notice retinopathy in its early stages. When detected early by your doctor, retinopathy can be successfully treated and additional vision loss prevented. Even so, it is so common among people with diabetes that after 15 years of the disease, 80 per cent of people not using insulin and 97 per cent of those using insulin show some signs of retinopathy.

The two major types of retinopathy are nonproliferative, sometimes called background retinopathy, and proliferative retinopathy. In the former, vessels in the eye close off or weaken and leak blood, fluid and fat into the eye, blurring vision. This blurred vision can cause blindness if the leaking occurs in the area of the retina near the optic nerve.

Less common, proliferative retinopathy occurs when new blood vessels grow in the retina. These new vessels grow differently than the original ones, rupturing easily. These ruptures are especially common if you also have high blood pressure. Following ruptures, blood may leak into the fluid of the eye, blocking the retina and/or forming scar tissue. If the scar tissue shrinks, it can tear the retina.

TREATMENT

Prevention and early detection are the most effective solutions, so people with diabetes should have their eyes checked annually after age 30. Working to keep your blood glucose levels at or near normal can also minimize the occurrence of this condition.

A blood pressure drug, lisinopril, has been shown to slow the progression of diabetic retinopathy. A British study of 530 people with type 1 diabetes published in the *Lancet* found that at the end of the two-year study, those who had received lisinopril had a slower progression of retinopathy as well as lower blood sugar levels.

Laser treatments can be used to burn away abnormal blood vessels, cauterize leaking ones and impede the development of new vessels. For some people vitrectomy – a procedure that removes excess blood and scar tissue, stops bleeding and replaces some of the substances within the eye while also repairing the damaged retina – can be helpful.

Retinopathy has a tendency to progress during pregnancy, especially if the mother has had diabetes for 10 years or more. Laser surgery may be helpful; even so, some vision may be lost.

Glaucoma

Glaucoma is nearly 1.5 times more common among people with diabetes than among those who don't have the disease. It is a condition in which the pressure of the fluid in the eye is so abnormally high that it compresses and obstructs the vessels supplying blood as well as the optic nerve.

Symptoms include dull, severe aching in and above the eye, fogged vision and the perception of rainbow rings around lights. Nausea and vomiting can occur, and the eye can become red and have a partly dilated pupil and a hazy cornea. The most common type – open-angle glaucoma – rarely occurs before age 40; closed-angle glaucoma has a rapid onset caused by a sudden, complete blockage of the outflow of fluid from the eye.

TREATMENT

Early detection through testing by an eye specialist is the key because the damage done by mounting pressure in the eye can be irreversible. Open-angle glaucoma can be treated with drops which reduce the pressure in the eye. In more resistant cases, oral medication may be used. A third, more aggressive technique which uses laser surgery to open the drainage channel in the eye or to create an artificial channel may be needed. Closed-angle glaucoma is a medical emergency that demands immediate attention in order to save vision.

Cataract

Cataracts are 1.6 times more likely among people with diabetes than in those who don't have the disease. Among people with diabetes they also occur at a younger age and progress much more rapidly. A cataract occurs because of a loss of transparency in the lens of the eye. This loss of transparency, over time, limits visual acuity and causes colour perception to be dulled.

TREATMENT

Eye surgery to replace the clouded lens with a new, artificial one.

Macular Oedema

Resulting from obstruction of circulation to the macula, the central portion of the retina, this causes an overall decrease in visual acuity. As a result, swelling occurs and is often accompanied by a hard lipid accumulation in the macular that may form a ring shape.

Neovascularization

Occurring most often in the non-retinal areas of the eye such as the iris, it produces rubeosis iridis. It also may occur in the rear chamber of the eye, causing haemorrhagic glaucoma.

Nephropathy

After 15 years of having diabetes, 10 to 20 per cent of people with type 2 diabetes develop kidney disease. Diabetes is the leading cause of end-stage renal disease or kidney failure, accounting for about 40 per cent of new cases. Research at the University of Minnesota and Northwestern University Medical School found in a study of 332,544 men between the ages 35 and 57 that those with diabetes were at 13 times greater risk for advanced kidney failure.

Dietary changes may provide some answers. At the New England Medical Center in Boston, a review of five studies comparing people with low-protein diets with those of people who ate average amounts found that those on the low-protein diet had slower rates of kidney disease.

This finding assessed the low-protein diet as leading to a 44 per cent reduced risk for diabetic kidney disease and may hold clues for preventing this kidney failure. Before changing your diet, talk with your doctor to be sure that a low-protein diet, including cutting back on meat, dairy products, nuts and beans while eating mostly fruit, vegetables and grains, is right for you.

Acute Renal Failure
In end-stage renal disease, the kidneys fail to clear the blood of waste and toxins, so the patient goes into acute renal failure and depends on frequent kidney dialysis to survive.

TREATMENT
Dialysis or kidney transplant.

During pregnancy, renal complications may increase for the mother. The amount of protein spilled into the urine during pregnancy should be rigorously monitored. Bed rest is the best treatment; medications may also be necessary.

Neuropathy

The connection between diabetes and damage to the nervous system is clear, but not totally explained. While complications from diabetes generally do not impair the brain or spinal cord, they can alter, damage dramatically and even destroy other nervous system functions.

About 60 to 70 per cent of people with diabetes have mild to severe forms of nervous system damage ranging from loss of sensation in the hands and feet to impaired digestion and other complications. In its most severe form, disease can lead to amputation: More than half the lower limb amputations in the US occur among people with diabetes.

Distal Symmetric Polyneuropathy (DSP)

This condition can attack the nerves of disparate parts of the body. As a result, arms, hands, legs and/or feet may feel numb, have no sensitivity to heat and cold or have sharp stabbing pains or a 'pins and needles' sensation; sufferers also report being unable to sense without looking how these body parts are positioned at any one time.

TREATMENT

Rigorous blood glucose control can help these symptoms disappear, although it may take as long as a year. Therapeutic exercise and the use of some medications can help, as can the topical application of over-the-counter products containing capsaicin, an extract of hot peppers.

A study in Sweden involving 126 adults with diabetes and neuropathy found that Mexiletine, a drug commonly prescribed to steady irregular heart beats, may relieve nerve pain. Study participants who received Mexiletine reported significantly less pain, and also that they were able to sleep better as a result.

Focal Neuropathy

Less common than DSP, focal neuropathy occurs as a result of damage to a single nerve or group of nerves. In general it can cause symptoms such as tingling, burning and numbness. Carpal tunnel syndrome is a typical example. This occurs when the median nerve of the forearm is squeezed by the carpal bones of the wrist.

TREATMENT

Once good blood glucose control is achieved, symptoms usually disappear.

Autonomic Neuropathy

Since the autonomic nervous system controls involuntary activities of the body – digestion, sexual response, sweating – it can be particularly perplexing. Damage to the autonomic nervous system can have disparate and challenging results for people with type 2 diabetes.

Gastrointestinal Upset

Ranging from nausea and vomiting to constipation and diarrhoea, this diabetes complication can slow the action of the stomach or gut muscles. This slowing can lead to inefficient emptying of the digestive track.

TREATMENT

Changes in eating habits such as smaller meals or more frequent meals, or just eating soft or liquid foods may help. With the help of your doctor you may need to experiment in order to find a diet that will work for you. Medications can help to promote stomach emptying and minimize constipation and diarrhoea.

Diabetic Gastroparesis

Occurring because the stomach empties slowly, diabetic gastroparesis is caused by damage to the nerves that control the pace at which food leaves the stomach and is digested. Symptoms include nausea and/or a sensation of fullness. The result for people with diabetes is that if food empties slowly and you have taken insulin before the meal, blood sugar may fall before the food has had a chance to be absorbed. Diets high in fat and fibre tend to worsen gastroparesis.

TREATMENT

Work with your doctor to find a solution that's right for you. You may need to take insulin right at the start of a meal to prevent low blood sugar and to match carbohydrate absorption. Medication may also be used to speed digestion.

Incontinence

When nerves to the bladder are damaged, muscles may become weak. As a result, bladder distention, leaking or difficulty emptying the bladder may occur.

TREATMENT

Bladder control training in which the bladder is emptied to a regular schedule can help. Some people find applying external pressure to the bladder aids in completely emptying the bladder; this can forestall leakage. Men may find it helpful to urinate seated. Medication, a catheter or surgery are other alternatives.

Loss of Sexual Function or Desire

Sexual function is a complex topic. For instance, sexual arousal for men and women is essentially a function of the autonomic nervous system. Yet the ability to engage in and enjoy intercourse is a complicated marriage of mental, emotional and physical capabilities.

For people with diabetes, this intricate process is compounded by the impact complications can have on the various systems and responses that make sexual activity possible and enjoyable. Men with diabetes develop impotence 10 to 15 years earlier than those without the disease; 50 to 60 per cent of all men over 50 who have diabetes also have some problems with impotence.

Some 35 per cent of women with diabetes experience some level of diminished sexual function, including reduced sensitivity to touch due to nerve damage as well as poor bladder control, vaginal dryness or reduced sexual desire.

The pressures of managing diabetes and fear of experiencing hypoglycaemia after intercourse can put a damper on a love-making session. So, too, can the loss of self-esteem which diabetes and its complications bring such as amputations, infection, incontinence and slow-healing injuries.

TREATMENT

Keeping blood glucose levels as close to normal as possible is important in preserving your ability to enjoy and engage in sexual activities. If you are experiencing sexual difficulties, talk with your doctor about suggestions and solutions. Don't assume you have to learn to live with this complication.

For women, lost sensitivity may be addressed with changes in sexual technique, or a more gentle or creative approach on the part of the partner. Vaginal dryness may be due to hormonal issues which may or may not be related to diabetes. Talk with your gynaecologist about hormonal testing to ensure your levels are where they should be. He or she may suggest hormone replacement therapy or over-the-counter lubricants. Poor bladder control may be resolved by emptying your bladder 30 minutes before sex. This has the added benefit of reducing the occurrence of bladder infections. Women with diabetes frequently experience vaginal Candida (thrush) infections. The chances of this happening can be reduced by using a small amount of antifungal cream as a lubricant. Boston University is studying whether the new medication for impotence, Viagra, may also help women reclaim their sexuality.

Impotence in men may be the result of blood vessel damage or nerve disease. A urologist can help diagnose the specific causes and recommend medication, therapies or surgery that can restore full or partial function. New medications such as Viagra, as well as apomorphine, may be appropriate for some men.

Men can help to preserve their ability to perform sexually by stopping smoking, decreasing alcohol intake and exercising regularly.

Some medications for high blood pressure control can contribute to impotence, as can drugs for depression, anxiety and peptic ulcers. If you use these medications, talk with your doctor about changes in dosage or other approaches to reduce their interference with sexual performance. Never stop taking medications without first talking with your doctor.

Postural Dizziness

When you exercise you may find your blood pressure rises quickly: Rising from a seated position or standing for long periods may leave you feeling light-headed or dizzy.

TREATMENT

This complication seems to be related to damage to the heart and blood vessels. If you drink alcohol, this too can cause blood pressure variations. Such medications as antidepressants, diuretics, nitroglycerine and some calcium-blocking drugs can also contribute. If you take these, talk with your doctor about ways in which dosages may be adjusted to minimize their effect on postural dizziness.

Other strategies your doctor may recommend include medication for low blood pressure or taking a proactive approach such as rising from a seated position slowly or avoiding standing for long periods of time. Wearing support stockings to prevent blood from pooling in your legs may also be a solution. If postural dizziness occurs most often upon rising from bed, try raising the head of the bed 6 to 9 inches. This slight elevation may help bridge the gap from being in a prone position to standing.

Gustatory Sweating

While eating, some people with diabetes break out in a sweat or experience flushing of the neck and chest.

TREATMENT

The exact cause of gustatory sweating isn't known, but it's believed to be related to having had high blood glucose levels for long periods of time. It appears to be brought on by eating cheese or chocolate as well as salty foods such as alcohol, fresh fruit, vinegar or anything pickled, so avoiding these also may limit or eliminate gustatory sweating. Significant improvement can be achieved by keeping your blood glucose levels in check.

Your doctor may recommend medications. One medication developed in the UK contains glycopyrrolate. British doctors tested a cream containing this substance on 13 people with diabetes. After using the cream, test subjects found sweating was stopped or significantly reduced.

Charcot's Joints

Most often affecting weight-bearing joints such as the ankles, Charcot's usually starts with bone thinning and a loss of feeling, especially in the extremities. Since bones are weakened and feeling is lacking, breaks may occur but go unnoticed, despite swelling. (Charcot's is sometimes misdiagnosed as osteoarthritis [OA], but Charcot's progress much more rapid than that of OA.) Because there's little or no pain due to neuropathy, you continue to use the affected joint. The injury worsens, muscles shrink and the joint may become deformed.

TREATMENT

Good blood glucose control can help prevent this condition. Early diagnosis is key in alleviating serious, permanent injury to joints. Treatment may include the use of splints to prevent additional joint deformity.

Foot Drop

This is a condition in which the foot cannot be raised properly due to damage to the nerves that supply the foot. The foot hangs limply from the ankle joint, causing it to catch as you walk.

TREATMENT

Your doctor may recommend wearing a splint, brace or other support to keep the foot in a proper position for walking.

Finger Stiffness

Because of high blood sugar over long periods of time, the tissue around the joints of your fingers may become stiff. This stiffness can keep you from straightening your fingers and make it hard for you to write, dress yourself, tie your shoes or pick up small objects.

To test yourself for this, hold your hands together with the palms facing and pressing together. If there's space between your right and left hands, you may have 'positive prayer sign', another diabetes complication.

TREATMENT

Keeping your blood glucose levels at or near normal is the best treatment.

Memory Loss

Poor blood glucose control appears to contribute to memory loss. An analysis of 19 diabetes studies seems to indicate that loss of memory and other cognitive function may be related to diabetes. The ability to remember words appears to be the most affected.

TREATMENT

Good blood glucose control appears to be the most effective treatment in preserving memory.

Other diabetic neuropathic conditions include:

- muscle wasting (atrophy) and paralysis as the result of nerve damage
- paralysis of the eye muscles so that the eyes no longer track in unison
- hammer toes, named for the distinctive, hammer-shaped position the toes take on. This complication occurs due to changes in muscle or in bony structures. It exposes the metatarsal heads of the feet, creating a greater potential for trauma and ulcer formation.
- Rocker foot occurs as bony structures continue to change and a loss of sensation in the ankle leads to a relaxation of

supporting tissues. The ankle becomes inflamed and the foot shortens and widens while the arch flattens.

Dental

Aggressive periodontal disease, gum-line cavities, dry mouth and infection are common among people with type 2 diabetes.

TREATMENT

Combined with rigorous blood glucose control, energetic daily dental hygiene can help prevent or minimize dental complications for patients with type 2 diabetes. You should also see your dentist twice a year for regular exams and cleanings to prevent plaque build-up, which can increase the likelihood of gum disease.

Dental Duty

Protect your teeth and gums with good glucose control plus these dental hygiene basics:

- Brush your teeth twice a day with a soft nylon brush that has rounded ends on the bristles. Also brush the rough upper surface of your tongue.
- Floss daily.

Get in touch with your dentist if:

- your gums are red, swollen or tender, bleed when you brush your teeth, or appear to be pulling away from your teeth
- you find pus between your teeth and gums when you touch the gums
- the fit of your dentures or partial plate changes or if your bite changes
- you have a persistent bad taste in your mouth or bad breath
- you have a consistent sore or 'raw' throat.

Infection

Type 2 diabetes can put you at risk for a wide variety of infections. That's because an excess of glucose makes the immune cells less effective. And many infection-causing agents – bacteria, viruses and fungi – feed on excess glucose.

Damaged nerves and reduced circulation can further complicate and increase the risk of infection. Damaged nerves lack sensation, so injuries may go unnoticed, creating opportunities for infection to set in and progress without immediate detection. Reduced circulation means people with diabetes take longer than people who don't have diabetes to heal from injuries, whether the injury is an ingrown toenail or a major surgical incision.

Urinary Tract Infections
Because of damage to the nerves, incomplete bladder emptying may occur. This unemptied urine is fertile ground for bacteria growth, resulting in frequent bladder infections.

TREATMENT
Emptying the bladder to a regular schedule and using external pressure to facilitate emptying can help. If bladder infections persist, your doctor may recommend antibiotics or solutions such as a catheter.

Vaginal Yeast Infections
The yeast *Candida albicans* which naturally inhabits a woman's vaginal tract thrives and over-multiplies where the environment is moist and glucose is abundant. It's a combination of conditions some women with diabetes provide.

TREATMENT
While a yeast infection can be very uncomfortable, with burning and itching that may be accompanied by a white discharge, it's usually not a danger to your health. To reduce the chances of developing a yeast infection:

- Wear breathable, all-cotton underwear and change it daily.
- Avoid panty hose and trousers that are tight and restrict air flow.

Your gynaecologist can recommend medications to relieve symptoms and eliminate the infection. The best long-term solution is good blood glucose control.

Foot Ulcers and Ulcerations

More likely to occur when circulation to the foot is poor and blood glucose levels uncontrolled, foot ulcers and ulcerations occur when a minor injury goes untreated. Infections set in, rapidly destroying layers of skin and other tissues and even reaching and infecting the bone.

Indiana University School of Medicine researchers found in a study of 352 people with type 2 diabetes that foot ulcers were most common among those over 40 who were also overweight and had neuropathy in their legs, leading to an inability to sense temperature changes. Other predictors include:

- dry and/or cracked foot skin
- poorly trimmed toenails
- fungal foot and nail infections.

TREATMENT

Good blood glucose control and diligent foot care are still the best ways to prevent foot ulcers. At the same time, doctors have recently determined treating ulcers with collagen may facilitate healing. In shallow ulcers, the collagen is applied topically. In deep ulcers, the wound must first be cleared surgically of dead and dying tissue, usually under local anaesthetic. Then collagen, plus antibiotics to control infection, are packed into the wound.

Other foot ulcer treatments showing promise are:

- Dermagraft, a bioabsorbable mesh onto which special skin cells from newborns have been added
- electric stimulation that activates the nerves and increases the blood flow.

In extreme cases, surgery may be necessary to remove the infected area; amputation is a possibility.

Gangrene

An advanced stage of infection, gangrene means affected tissues have died, usually as the result of infection or lost blood supply. This is a serious medical emergency which can occur when a person with diabetes fails to seek proper care for an injury.

TREATMENT

Without proper medical care, usually antibiotics and surgical removal of affected tissues, a gangrene infection can spread to invade other tissues. In some cases, amputation is the only way to remove diseased tissue and save the patient's life.

Staphylococcal Infections

These occur as itchy spots on the buttocks, knees and elbows.

TREATMENT

See your doctor, who may recommend topical treatment to eliminate the infection and relieve symptoms.

Good Foot Care

If you have diabetes, restricted circulation and nerve damage can make your feet vulnerable to a wide variety of conditions and to injuries. The injured feet of a person with diabetes heal slowly, and this creates a window for infection. To limit damage and injury to your feet, remember:

- Inspect your feet daily for blisters, cuts, scratches and tenderness. If necessary, use a mirror to examine thoroughly the underside of the feet.
- Wash your feet thoroughly at least once a day. Dry carefully and completely between the toes.
- Inspect the insides of shoes daily for objects and torn or rough areas; be sure your shoes contain plenty of room in the toe box and fit comfortably in the heel. Consider having your shoes custom-fitted.
- Wear only shoes made of breathable materials; change shoes at least daily.
- Avoid temperature extremes by testing bath water before stepping in.
- Use lotion on your feet but avoid getting it between the toes; your doctor may recommend over-the-counter or prescription lotions to ease dryness and the build-up of calluses.
- Do not use heating pads or hot water bottles, and avoid soaking your feet in hot water.
- Cut nails in contour with the toes; better yet, have them trimmed by a podiatrist.
- Don't use chemical plasters or ointments to remove corns or calluses; do not cut corns or calluses away.
- Do not use adhesive tape on the feet, as it can irritate the skin.
- Never walk barefoot or wear open-toed shoes.
- If your feet have decreased sensation, break in shoes slowly, wearing different pairs throughout the day.
- Notify your doctor immediately if you find a blister, crack, cut or other inflammation on your feet.

Mental Health

Living with diabetes can be stressful. For people with diabetes, everyday decisions can be crucial ones of continued well-being and long-term health. While the occurrence of depression is the same for people with and without diabetes, people with diabetes may find themselves grappling with depression for reasons that are specific to their disease.

The links between depression and diabetes generally relate to concerns about:

- long-term physical health
- financial demands of diabetes as well as barriers the disease may present to work
- changes in relationships with family and friends, employer and activities you love best as a result of having to cope with diabetes.

Depending upon your ability to control your disease, you may be fearful of hypoglycaemia or hyperglycaemia. You may feel anxious about managing medication correctly. And if you are already experiencing complications such as retinopathy or sexual dysfunction, anxiety may increase exponentially.

If you're seriously depressed, the distress can actually alter your ability to manage your blood glucose levels. And it can contribute to such behaviours as inappropriate diet, alcohol use and failure to exercise. Depression can intensify the discomfort you may be feeling as a result of other complications.

TREATMENT

Talking with your doctor can help you gain the confidence and the knowledge you need to manage your disease successfully. This confidence, when combined with the help of a support group, may aid you in achieving more stable and hopeful feelings about your life.

Exercise can be especially helpful in relieving symptoms of depression and anxiety. That's good news, because regular, doctor-approved exercise can play a key role in diabetes management, weight loss and the reduction of complications.

An Indiana University study found that just 20 minutes of leg exercises led to a drop in anxiety levels, even when the exercises were done in a mild way and at a light pace. The researchers further found that just 5 minutes of exercise led to a reduction in anxiety levels which could still be felt two hours later. They speculate that longer, more intense exercise sessions could have greater benefits.

Medication and/or therapy also may be options. Three studies in the 1997 issue of *Psychological Annals, Psychosomatic Medicine and Diabetes* have shown people with diabetes respond equally well to antidepressant medication and to therapy.

Symptoms of Depression

If you experience three or more of the following symptoms for two weeks or longer, or if you are having suicidal thoughts, seek help from a mental health professional.

- feelings of sadness, hopelessness and extreme irritability
- loss of interest or pleasure in activities you previously enjoyed
- trouble sleeping, sleeping too much, or awakening at an inappropriate time and being unable to go back to sleep
- weight loss or weight gain
- difficulty concentrating
- extreme tiredness or loss of energy
- feelings of anxiety or guilt.

Pancreatic Cancer

Researchers at the National Institutes of Diabetes and Digestive and Kidney Diseases (NIDDK) in the US have reported a finding that may indicate a connection between diabetes and pancreatic cancer.

NIDDK researchers reported this potential connection based on a review of 20 studies done between 1975 and 1994 which dealt with pancreatic cancer risk factors. In 18 of the 20 studies, diabetes emerged as a risk factor.

TREATMENT

Good blood glucose control and a healthy lifestyle remain the strongest lines of defence against diabetic complications.

Pregnancy and Diabetes Complications

Whether you had diabetes before you became pregnant or developed gestational diabetes, staying healthy and having a healthy baby present particular challenges to women with diabetes.

Women with diabetes who want to become pregnant should talk with their doctor or a midwife about the steps they can take to minimize complications. For a woman whose pregnancy precedes her diabetes, there are many challenges.

The pregnant woman with diabetes who already suffers from diabetic complications may increase the severity of them and must manage all aspects of her disease rigorously during pregnancy. In some cases your doctor may recommend avoiding pregnancy because it could put your health at risk, particularly if you already have cardiovascular disease, kidney failure or serious gastrointestinal neuropathy.

The pregnant woman with diabetes is at increased risk for high blood pressure.

Birth defects are another complication whose potential should be carefully assessed before pregnancy. The average risk of birth defects is 2 to 3 per cent for healthy women; for babies whose mothers have diabetes and maintain good blood glucose control, the risk is 6 to 12 per cent. Typical defects include central nervous system abnormalities as well as heart and kidney defects.

Gestational Diabetes

Gestational diabetes is usually a temporary form of diabetes which occurs when the hormones associated with pregnancy create extra insulin resistance. It's most often identified about halfway through pregnancy. Risk factors for gestational diabetes include:

- a family history of diabetes
- a history of gestational diabetes in previous pregnancies
- a history of glucose intolerance outside of pregnancy
- obesity
- patient's own birthright having been greater than nine pounds
- poor obstetrics history, such as habitual abortions or a previous baby with a birth weight greater than nine pounds
- unexplained stillbirths
- birth defects in other children
- history of toxaemia
- recurrent urinary tract infections
- excess amniotic fluid.

Being overweight appears to be a significant contributing factor; a pregnant woman who is overweight has a three in four chance of developing gestational diabetes.

True gestational diabetes disappears after delivery. Women who have had gestational diabetes are at greater risk for developing type 2 diabetes 5 to 15 years after pregnancy.

A small number of women continue to have diabetes after delivery. They are generally believed to be people who were previously undiagnosed and whose disease was made evident by pregnancy.

FIGURE 9.2

Recommendations for Care of Type 2 Diabetes

Yearly (in addition to the quarterly recommendations)

Complete physical examination.

Complete eye examination, with dilitation by an ophthalmologist.

Blood chemistry analysis with a fasting lipid profile.

Urinary testing for microalbuminuria.

If cardiac disease present – electrocardiogram.

Quarterly

    Blood pressure check, pulse, heart rate.

    Foot examination by a podiatrist.

    HbA1c (glycohaemoglobin) measurement and evaluation.

    Review of self glucose monitoring (SGB) results.

Monthly

    Self-review of dietary management.

    Glycoprotein analysis (fructosamine testing).

    Feet inspection for sores, cracks, potential problems.

    Urinary testing for protein.

    Review of lifestyle goals.

    Modification of diabetes cure plan if needed.

Finally, the Good News

Just knowing about the complications of diabetes can be depressing. Diabetes at its worst is among the most devastating of all diseases, namely because it has the ability to exploit a person's physical weaknesses, no matter what they are.

How could there possibly be any good news about diabetes?

Well, there is, and it is this: Type 2 diabetes is a disease of lifestyle that can be reversed by adopting healthy habits. You don't have to be a victim of this terrible disease if you take control and decide you are going to beat it.

Yes, diabetes is beatable and with this book you have the tools to do just that. As I tell my patients, 'I can give you the roadmap to wellness, but you have to drive the car.'

Drive carefully.

CHAPTER TEN

The Future
is Now

The discovery in 1921 that the injection of secretions from a portion of the pancreas – the islets of Langerhans – could relieve symptoms of diabetes changed lives. Since then, research has developed new techniques and treatments that improve the lives of people with diabetes, even as understanding of the mechanisms and side-effects of type 2 diabetes continues to emerge.

Despite all the new and improved methods of treating type 2 diabetes, prevention still stands out as the single most effective way to avoid or limit diabetic complications. For those who have type 2 diabetes, self-monitoring and serious attention to lifestyle changes still stand as the best ways to control, slow and even reverse the course of diabetes. Most cases of type 2 diabetes can be cured, but it takes serious attention to all aspects of lifestyle which caused the diabetes to begin with. By following the nine steps outlined in the Diabetes Cure, you are doing just that, attacking this insidious disease on all fronts.

The essential role of prevention and self-monitoring was crystallized via the Diabetes Control and Complications Trial (DCCT) in the US. Completed in 1988, this study confirmed that the risk of diabetic complications could be reduced with tight control to keep blood glucose levels as close to normal as possible. This study confirmed what many had previously thought, and further reinforced the roles played by prevention and individual responsibility in living a long and fulfilling life, despite diabetes. At the same time, results from the DCCT continue to play a part in the development of individualized approaches to diabetes management.

Beyond Prevention

The numbers of people with type 2 diabetes have risen 700 per cent since the 1950s. This increase reflects a number of emerging trends, including the increase in average age – the most common age at which diabetes is diagnosed is 51.

Other factors include growth among ethnic populations, who appear to be at higher risk for diabetes, as well as increases in weight and decreases in levels of activity across the population. These trends have led to the development and exploration of new and diverse treatments.

Medication

In the past 10 years, a number of new medications have emerged. In addition to using HCA and following the steps outlined in the Diabetes Cure, you might want to discuss some of these medications with your doctor. This is especially true if you are insulin-dependent. *Family Practice News* reported in 1996 the use of a combination of drugs to treat people with insulin-dependent type 2 diabetes leads to reduced side-effects and fewer long-term complications than a single drug.

- The blood pressure drug lisinopril slows the development of diabetic retinopathy which occurs in 70 per cent of people with type 2 diabetes.
- The cholesterol-lowering drug simvastatin is in the newest class of medications for this purpose. In a Swedish study reported in *Diabetes Care*, people with diabetes who used simvastatin experienced a 55 per cent decrease in their risk of

heart attack and a 43 per cent increase in death rate due to heart attack.

- The ACE inhibitor fosinopril appears to create a 50 per cent lowered risk of stroke, heart attack and chest pain among people with diabetes and high blood pressure who use the drug.
- Mexiletine, a cardiac drug, is now being studied for use in treating people with diabetes who have a painful nerve disease as a result of their condition.
- Tagamet, a medication available since the early 1980s for ulcers and heartburn, is showing promise for people with type 2 diabetes. When given in a liquid form, researchers at the University of Oslo in Norway found Tagamet helped test subjects shed weight, lower blood pressure and control blood glucose levels.
- Diabetic kidney disease is the leading cause of people undergoing kidney dialysis. A new test has been developed that detects kidney disease early, allowing for earlier detection and treatment. In combination with ACE inhibitors, this early detection can limit the occurrence of kidney disease, reducing the need for dialysis.

Other Innovations

Although the goal of the Diabetes Cure is to cure diabetes or lessen its effects, there are people with type 2 diabetes who are using insulin and are frustrated at having to use needle and syringe to administer it. For them, a method of needle-free insulin administration is in development. Sometimes called 'the holy grail' of diabetes care, it allows people with diabetes to inhale insulin via microscopic particles of crystallized insulin placed inside a plastic tube. Once the particle pack opens, the vapour can be inhaled in one breath. The insulin is absorbed instantly into the bloodstream through the vast surface of the lungs.

Preliminary research is also underway in the development of an artificial pancreas. Unlike the insulin pump, which is worn on the outside of the body and whose use is generally limited to type 1 diabetes, this device would be implanted inside the body. It would include the insertion of an internal catheter into the abdominal cavity to convey insulin, plus a built-in glucose sensor and sensitive systems that would automatically detect when the body needs more insulin and automatically release it.

It is good to know about these technological advances, but frankly I hope you never have to use them. Diabetes can be beaten, and the Diabetes Cure provides all of the necessary ingredients for success.

When patients ask me about future medical developments, hoping that a future diabetes medication will have no side-effects and will allow them to live their lives unchanged, I tell them that the future is now.

Without diligent change on your part, modern medicine can only slow the progress of your disease. It cannot cure it.

If this information seems depressing, let me tell you about one of my worse-off patients, Maxine.

Maxine illustrates a more severe case of type 2 Diabetes. At 5' 1", Maxine had been placed on insulin for her diabetes about two years prior to seeing me. Recently, she had moved to the area and was establishing new relationships and finding a new doctor, dentist and pharmacy.

The move was emotionally trying and had left her feeling depressed and fatigued. She felt that her Prozac was not helping her as it used to, and she was frustrated with herself for the short temper she was exhibiting towards her family. Biscuits seemed to help with some of her anxiety, but only temporarily. Maxine wanted help.

Her glucose was 'out of control' at 305. Never had it been so consistently high. She had tried several times to lose weight during the previous two months, but her scales fluctuated by just a few pounds. Determined to get back on track, she began using HCA in addition to the insulin. She noticed that the HCA curbed her appetite, and she felt fuller, faster. Her weight began to drop. After eight weeks she had lost more than 10 pounds. Her fasting glucose dropped with her weight, down to 211. With the drop in weight and the glucose level came a rise in energy.

With a more stable glucose level and body weight, Maxine began to question her endocrinologist about what she needed to do to 'get off the insulin'. He assured her that if she continued to increase her activity, lose weight and take HCA with chromium, she would be on the road to stopping her need for insulin altogether.

She commented to her doctor that, 'The Prozac seems to be working better' and acknowledged that all of her was working better. She reported less frustration, less irritability and more patience in her new surroundings.

Although still seriously overweight, Maxine had realized the importance of focusing on her health. She has now become an advocate of 'making time for health', and as a result has more time and energy for other things.

You have in your hands the same guidebook that Maxine and others have used. It is one that can lead you out of the dangerous terrain defined by the disease known as type 2 diabetes. Whether you have diabetes, or are at risk of getting it, now is the time to use the information in this book to put your health in order.

Useful References

Medical Literature

Alterman, Seymour L., *How to Control Diabetes* (NY: Ballantine Books, 1996)

American Diabetes Association, 'Clinical Practice Recommendations 1998', Introduction and Position Statement, vol. 21, supplement 1.

Anderson, G. H., 'Regulation of Food Intake', in Maurice E. Shils, James A. Olson and Moshe Shike (eds), *Modern Nutrition in Health and Disease* (8th edn; Philadelphia: Lea and Febiger, 1994): 524—36

Barth, C. J. Hackenschmidt, H. Ullman and K. Decker, 'Inhibition of Cholesterol Syntheses by (-)-Hydroxycitrate in Perfused Rat Liver. Evidence for an Extramitochondrial Mevalonate Synthesis from Acetyl Coenzyme A', *Archives of Biochemistry and Biophysics* 22.3 (May 1972): 343—6

Berkhout, T. A., 'The Effect of (-)-Hydroxycitrate on the Activity of the Low-Density-Lipoprotein Receptor and 3-hydroxy-3-methylglutaryl-CoA Reductase Levels in the Human Hepatoma Cell Line Hep G2', *Biochemistry Journal* 272 (1990): 181—6

Boll, M., E. Sorensen and E. Balieu, 'Naturally Occurring Lactones and Lactames. III. The Absolute Configuration of the Hydroxycitric Acid Lactones: Hibiscus Acid and Garcinia Acid', *Acta Chemica Scandinavica* 23 (1969): 283—93

Bonadonna, R. G. and R. A. Defronzo, 'Glucose Metabolism in Obesity and Type II Diabetes', in Per Bjorntorp and Bernard N. Brodoff (eds), *Obesity* (Philadelphia: J. B. Lippincott, 1992): 474—6

Bonora, E., C. Coscelli and U. Butturini, 'Relationship Between Insulin Resistance, Insulin Secretion and Insulin Metabolism in Simple Obesity', *International Journal of Obesity* 9 (1985): 307—12

Bowden, C. R., K. D. White and G. F. Tutwiler, 'Energy Intake of Cafeteria-Diet and Chow-Fed Rats in Response to Amphetamine, Fenfluramine, (-)-Hydroxycitrate, and Naloxone', *Journal of Obesity and Weight Regulation* 4.1 (Spring 1985): 5—13

Brunengraber, H., J. R. Sabine, M. Boutry and J. M. Lowenstein, '3-beta-Hydroxysterol Synthesis by the Liver', *Archives of Biochemistry and Biophysics* 150 (1972): 392—6

Chee, H., D. R. Romsos and G. A. Leveille, 'Influence of (-)-Hydroxycitrate on Lipogenesis in Chickens and Rats', *Journal of Nutrition* 107.1 (1977): 112—19

Cheema-Dhadli, S., M. L. Halperin and C. C. Leznoff, 'Inhibition of Enzymes Which Interact with Citrate by (-)-Hydroxycitrate and 1,2,3,-Tricarboxybenzene', *European Journal of Biochemistry* 38 (1973): 98—102

Clouatre, D., *Anti-Fat Nutrients* (2nd edn; San Francisco: PAX Publishing, 1993)

Clouatre, D. and M. Rosenbaum, 'The Diet and Health Benefits of HCA (Hydroxycitric Acid)', (New Canaan, CT: Keats Publishing, 1994)

Conte, A. A., 'A Non-Prescription Alternative in Weight Reduction Therapy', cited in A. Conte and M. Majeed, *A Revolutionary Herbal Approach to Weight Management* (chapter 6; Burlingame, CA: New Editions Publishing, 1994)

Danforth, M., 'Weight Loss Clinical Outcome Study', Legere Pharmaceuticals Protocol Using All Natural Agents, June 22 1998. Abstract, via Legere

Eastman, R. C., 'Prevention of Type 2 Diabetes', *Current Approaches to the Management of Type-2 Diabetes: A Practical Monograph*, National Diabetes Education Initiative for Health Care Professionals, Professional Graduate Services, 1997: 43—6

Finney, Lois, R. D. Schmidt and Michael Gonzalez-Campoy, 'Dietary Chromium and Diabetes: Is There a Relationship?', *Clinical Diabetes* 15.1 (January—February 1997): 6—8

Useful References

Frawley, David, *The Yoga of Herbs* (Santa Fe, NM: Lotus Press, 1986)

Fujimoto, M. D., 'A National Multicenter Study to Learn Whether Type II Diabetes Can Be Prevented: The Diabetes Prevention Program', *Clinical Diabetes* 15.1 (January—February 1997): 13—15

Garber, Alan J., 'For Better or for Worse', *Clinical Diabetes* 15.1 (January—February 1997): 2—3

—, 'When Does Diabetes Begin, and How Do We Know It?', *Clinical Diabetes* (July—August 1997): 146—7

Gavin, James R. III and Hal Singer, 'Putting Findings to Work', *Patient Care* February 15, 1998: 13—14

Gibbons, G. F., 'The Metabolic Route by Which Oleate *is* Converted into Cholesterol in Rat Hepatocytes', *Biochemistry Journal* 235 (1986): 19—24

Greenwood, M. R. C., M. P. Cleary, R. Gruen, D. Blasé, J. S. Stern, J. Triscari and A. C. Sullivan, 'Effect of (-)-Hydroxycitrate on Development of Obesity in the Zucker Obese Rat', *American Journal of Physiology* 240 (1981): E72—8

Guyton, A. C., chapter 67 in *Textbook of Modern Physiology* (8th edn; Philadelphia: W. B. Saunders, 1991

Hirsch, I. B., 'Standards of Care for the Treatment of Diabetes', *Clinical Diabetes* 16.1 (1998): 26—9

Hobbs, L. S., '(-)-Hydroxycitrate (HCA)', in *The New Diet Pills* (Irvine, CA: Pragmatic Press, 1994): 161—74

Hoffmann, C. E., H. Andres, L. Weiss, C. Kreisel and R. Sander, 'Lipogenesis in Man: Properties and Organ Distribution of ATF Citrate (pro-3S)-Lyase', *Biochimica et Biophysica Acta* 620 (1980): 151—8

Horton, E. S., 'Therapeutic Options for the Treatment of Type 2 Diabetes', *Current Approaches to the Management of Type-2 Diabetes: A Practical Monograph*, National Diabetes Education Initiative for Health Care Professionals, Professional Graduate Services, 1997: 35—42

Jaedig, S. and N. C. Henningsen, 'Increased Metabolic Rate in Obese Women After Indigestion of Potassium, Magnesium- and Phosphate-Enriched Orange Juice or Injection of Ephedrine', *International Journal of Obesity* 15 (1991): 429—36

Kaats, G. R., D. Pullin, L. K. Parker, P. L. Keith and S. Keith, 'Reductions of Body Fat as a Function of Taking a Dietary Supplement Containing Garcinia Cambogia Extract, Chromium Picolinate and L-Carnitine – A Double Blind Placebo Controlled Study' [abstract], *Age* 14 (1991): 138

King, Dana E., Virginia Peragallo-Dittko, William Polonosky, James Proshaska and Frank Vinicor, 'Strategies for Improving Self-Care', *Patient Care* February 15, 1998: 91—111

Kitabchi, Abbas E. and Michael Bryer-Ash, 'NIDDM: New Aspects of Management', *Hospital Practice* March 15, 1997: 135—64

Lean, M. E., 'Evidence for Brown Adipose Tissue in Humans', in Per Bjorntorp and Bernard Brodoff (eds), *Obesity* (Philadelphia: J. B. Lippincott, 1992)

Leon, Gloria R., 'Behavioral Approach to Obesity', *American Journal of Clinical Nutrition* 30 (1977): 785—9

Levin, B. E., J. Triscari and A. C. Sullivan, 'Altered Sympathetic Activity During Development of Diet-Induced Obesity in Rats', *American Journal of Physiology* 244.2 (1983): R347—55

Lewis, Y. S., 'Isolation and Properties of Hydroxycitric Acid', in John M Lowenstein (ed), *Citric Acid Cycle*, vol. 13 of *Methods of Enzymology* (NY: Academic Press, 1969): 613—19

Lowenstein, J. M., 'Effect of (-)-Hydroxycitrate on Fatty Acid Synthesis by Rat Liver in Vivo', *Journal of Biological Chemistry* 246.3 (February 10, 1971): 629—32

—, 'Experiments with (-)-Hydroxycitrate', in W. Bartley, H. L. Kornberg and J. R. Quayle (eds), *Essays in Cell Metabolism* (London: Wiley-Interscience, 1970): 153—66

Lowenstein, J. M. and H. Brunengraber, 'Hydroxycitrate', in John M. Lowenstein (ed), *Lipids*, vol. 72 of *Methods of Enzymology* (NY: Academic Press, 1981): 486—97

McCarty, M. F., 'Inhibition of Citrate Lyase May Aid Aerobic Endurance', *Medical Hypotheses* 45 (1995): 247—54

—, 'Optimizing Exercise for Fat Loss', *Medical Hypotheses* 44 (1995): 325—30

—, 'Promotion of Hepatic Lipid Oxidation and Gluconeogenesis as a Strategy for Appetite Control', *Medical Hypotheses* 42 (1994): 215—25

—, 'Reduction of Free Fatty Acids May Ameliorate Risk Factors Associated with Abdominal Obesity', *Medical Hypotheses* 44 (1995): 278—86

McCarty, M. F. and M. Majeed, 'The Pharmacology of Citrin', in M. Majeed *et al.* (eds), *Citrin: A Revolutionary Herbal Approach to Weight Management* (Burlingame, CA: New Editions Publishing, 1994): 34—51

Manchester, Jill, Alexander V. Skurat, Peter Roach, Stephen D. Hauschka and John C. Lawrence Jr., 'Increased Glycogen Accumulation in Transgenic Mice Overexpressing Glycogen Synthase in Skeletal Muscle', *Proceedings of the National Academy of Sciences* 93.20 (1996): 10707—11

Mathias, M. M., A. C. Sullivan and J. C. Hamilton, 'Fatty Acid and Cholesterol Synthesis from Specifically Labeled Leucine by Isolated Rat Hepatocytes', *Lipids* 16.10 (1981): 739—43

Nomura, F. and K. Chinishi, 'Liver Function in Moderate Obesity – Study in 534 Moderately Obese Subjects Among 4,613 Male Company Employees', *International Journal of Obesity* 10 (1986): 349—54

O'Connor, Deborah L., 'Is Optimal Glucose Control Worthwhile?@, *Patient Care* February 15, 1998: 23—4

Olefsky, J. M., *Current Approaches to the Management of Type-2 Diabetes: A Practical Monograph*, National Diabetes Education Initiative for Health Care Professionals, Professionals Graduate Services, 1997

—, 'Pathophysiology of Type-2 Diabetes', *Current Approaches to the Management of Type-2 Diabetes: A Practical Monograph*, National Diabetes Education Initiative for Health Care Professionals, Professionals Graduate Services, 1997: 7—14

Panksepp, J., A. Pollack, R. Meeker and A. C. Sullivan, '(-)-Hydroxycitrate and Conditioned Aversions', *Pharmacology Biochemistry and Behavior* 6 (1977): 683—7

Perry, L. M. and J. Metzger, *Medicinal Plants of East and Southeast Asia* (Cambridge, MA: MIT Press, 1980): 174—6

Pullinger, C. R. and G. F. Gibbons, 'The Role of Substrate Supply in the Regulation of Cholesterol Biosynthesis in Rat Hepatocytes',

Biochemistry Journal 210 (1983): 625—32

Rao, R. Nageswara and K. K. Sakariah, 'Lipid-Lowering and Antiobesity Effect of (-)-Hydroxycitric Acid', *Pergamon Journals*, Let., 1988. Abstract

Rath, M., *Eradicating Heart Disease* (San Francisco: Health Now, 1993)

Schardt, D. and S. Schmidt, 'Supplement Watch: Chromium', *Nutrition Action Healthletter* May 1996: 10—11

Scoville, B. A., 'Review of Amphetamine-like Drugs by the Food and Drug Administration: Clinical Data and Value Judgments', in *Obesity in Perspective: Proceedings of the Fogarty Conference* (Washington, D.C.: U.S. Government Printing Office, 1973): 441—3

Sener, A. and W. J. Malaisse, 'Hexose Metabolism in Pancreatic Islets. Effect of (-)-Hydroxycitrate upon Fatty Acid Synthesis and Insulin Release in Glucose-Stimulated Islets', *Biochimie* 73 (1991): 1287—90

Sergio, W., 'A Natural Food, the Malabar Tamarind, May Be Effective in the Treatment of Obesity', *Medical Hypotheses* 27 (1988): 39—40

Silverstone, T., 'Drugs, Appetite and Obesity: A Personal Odyssey', *International Journal of Obesity* 16, supplement 2 (1992): S49—52

Stock, M. J., 'Thermogenesis and Energy Balance', *International Journal of Obesity* 16, supplement 2 (1993): S13—16

Sullivan, A. C., 'The Influence of (-)-Hydroxycitrate on in Vivo Rates of Hepatic Glycogenesis, Lipogenesis and Cholesterolgenesis', *Federation Proceedings* 33 (1974): 656

—, 'Metabolic Inhibitors of Lipid Biosynthesis as Anti-obesity Agents', in P. B. Curtis-Prior (ed), *Biochemical Pharmacology of Obesity* (Amsterdam: Elsevier Science Publishers, 1983): 311—25

—, 'Reactivity and Inhibitor Potential of Hydroxycitrate Isomers with Citrate Synthase, Citrate Lyase and ATP Citrate Lyase', *Tile Journal of Biological Chemistry* 252.21 (1977): 7583—90

Sullivan, A. C. and K. Comai, 'Pharmacological Treatment of Obesity', *International Journal of Obesity* 2 (1978): 167—89

Sullivan, A. C. and R. K. Gruen, 'Mechanisms of Appetite Modulation by Drugs', *Federation Proceedings*, Part I (1985): 139—44

Sullivan, A. C., J. G. Hamilton, O. N. Miller and V. R. Wheatley, 'Inhibition of Lipogenesis in Rat Liver by (-)-Hydroxycitrate',

Archives of Biochemistry and Biophysics 150 (1972): 183—90. Abstract

Sullivan, A. C. and J. Triscari, 'Metabolic Regulation as a Control for Lipid Disorders. I. Influence of (-)-Hydroxycitrate on Experimentally Induced Obesity in Rodent', *American Journal of Clinical Nutrition* 30 (1977): 767—76. Abstract

—, 'Possible Interaction Between Metabolic Flux and Appetite', in D. Novin, W. Wyriwicka and G. Bray (eds), *Hunger: Basic Mechanisms and Clinical Implications* (NY: Raven Press, 1976): 115—25

Sullivan, A. C., J. Triscari and L. Cheng, 'Appetite Regulation by Drugs and Endogenous Substances', in M. Winick (ed), *Nutrition and Drugs – Current Concepts in Nutrition* (vol. 12; NY: John Wiley and Sons, 1983): 139—67

Sullivan, A. C., J. Triscari and K. Comai, 'Pharmacological Modulation of Lipid Metabolism for the Treatment of Obesity', *International Journal of Obesity* 8, supplement 1 (1984): 241—8

Sullivan, A. C., J. Triscari and J. G. Hamilton, 'Hypolipidemic Activity of (-)-Hydroxycitrate', *Lipids* 12.1 (1977): 1—9

Sullivan, A. C., J. Triscari, J. G. Hamilton and O. N. Miller, 'Effect of (-)-Hydroxycitrate upon the Accumulation of Lipid in the Rat: II. Appetite', *Lipids* 9 (1974): 129—34

Sullivan, A. C., J. Triscari, J. G. Hamilton, O. N. Miller and V. R. Wheatley, 'Effect of (-)-Hydroxycitrate upon the Accumulation of Lipid in the Rat: I. Lipogenesis', *Lipids* 9 (1974): 121—8

Sullivan, A. C., J. Triscari and H. E. Spiegel, 'Metabolic Regulation as a Control for Lipid Disorders. II. Influence of (-)-Hydroxycitrate upon Genetically and Experimentally Induced Hyper triglyceridemia in the Rat', *American Journal of Clinical Nutrition* 30 (1977): 777—84. Abstract

Thom, E., 'Hydroxycitrate (HCA) in the Treatment of Obesity', *International Journal of Obesity* 20, supplement 4 (1996): 75. Abstract

Triscari, J. and A. C. Sullivan, 'Comparative Effects of (-)-Hydroxycitrate and (+)-allo-Hydroxycitrate on Acetyl CoA Carboxylase and Fatty Acid and Cholesterol Synthesis in Vivo', *Lipids* 12.4 (April 1977): 357—63

Voet, D. and J. C. Voet, *Biochemistry* (NY: John Wiley and Sons, 1990)

Wadden, T. A. and A. J. Stunkard, 'Psychosocial Consequences of Obesity and Dieting', in *Obesity: Theory and Therapy* (2nd edn; NY: Raven Press, 1993)

Walsh D'epiro, Nancy, 'Alarming Diabetes Numbers from the CDC', *Patient Care*, February 15, 1998: 18

—, 'Bloodless Monitoring: Coming Soon?' *Patient Care*, February 15, 1998: 20

—, 'Of Lost Limbs', *Patient Care*, February 15, 1998: 19

Watson, J. A., M. Fang and J. M. Lowenstein, 'Tricarballylate and Hydroxycitrate: Substrate and Inhibitors of ATP: Citrate Oxaloacetate Lyase', *Archives of Biochemistry and Biophysics* 135 (1969): 209—17

The Wealth of India (Raw Materials) (vol. 4; New Delhi: Council Sci. Ind. Res., 1956): 99

Wheeler, Thomas J., 'Hydroxycitrate as a Weight Loss Ingredient', Department of Biochemistry, University of Louisville School of Medicine, 1996. Abstract, via Internet

Zinman, B., 'Guidelines for the Management of Type 2 Diabetes', *Current Approaches to the Management of Type-2 Diabetes: A Practical Monograph*, National Diabetes Education Initiative for Health Care Professionals, Professional Graduate Services, 1997: 19—22

Recommended Reading

Alterman, Seymour L., *How to Control Diabetes* (NY: Ballantine Books, 1996)

American Heart Association, *Your Heart: American Heart Association's Complete Guide to Heart Health* (NY: Simon and Schuster, 1995)

Balch, James F., *Prescription for Nutritional Healing* (2nd edn; NY: Avery Publishing Group, 1997)

Barnard, Neal, *Eat Right, Live Longer* (NY: Crown, 1995)

Braley, James, *Dr Braley's Food Allergy and Nutrition Revolution* (NY: Keats Publishing, 1992)

Carlson, Richard, *Don't Sweat the Small Stuff ... and It's All Small Stuff* (NY: Hyperion, 1997)

Chopra, Deepak, *Ageless Body, Timeless Mind* (NY: Harmony Books, 1993)

Covey, Stephen R., *First Things First* (NY: Simon and Schuster, 1994)

—, *The Seven Habits of Highly Effective People* (NY: Simon and Schuster, 1989)

Diamond, Harvey and Marilyn Diamond, *Fit for Life* (NY: Warner Books, 1985)

—, *Fit for Life II: Living Health* (NY: Warner Books, 1987)

Greene, Bob and Oprah Winfrey, *Make the Connection* (NY: New Sage Press, 1997)

Kamen, Betty, *The Chromium Connection: A Lesson in Nutrition* (Novato, CA: Nutrition Encounter, 1996)

Katahn, Martin, *The T-Factor Diet* (NY: W. W. Norton, 1993)

Kelley, David B., *American Diabetes Association Complete Guide to Diabetes* (NY: Bantam Books, 1996)

Klatz, Ronald and Robert Goldman, *Stopping the Clock: Dramatic Breakthroughs in Anti-Aging and Age Reversal Techniques* (NY: Keats Publishing, 1996)

Larkin, Marilyn, *Redux, the Revolutionary Weight Loss Drug* (NY: Avon Books, 1997)

Lyman, Howard F., *The Mad Cowboy: Plain Truth from the Cattle Rancher Who Won't Eat Meat* (NY: Scribner, 1998)

McDougall, John A., *The McDougall Program: Twelve Days to Dynamic Health* (NY: Penguin Group, 1990)

—, *The McDougall Program for Maximum Weight Loss* (NY: Penguin Group, 1994)

Marti, James E., *Alternative Health Medicine Encyclopedia: The Authoritative Guide to Holistic and Nontraditional Health Practices* (Detroit: Visible Ink Press, 1995)

Ornish, Dean, *Dr. Dean Ornish's Program for Reversing Heart Disease* (NY: Ballantine Books, 1996)

—, *Love and Survival: The Scientific Basis for the Healing Power of Intimacy* (NY: HarperCollins, 1997)

Pizzorno, Joseph, *Total Wellness: Improve Your Health by Understanding the Body's Healing Systems* (Rocklin, CA: Prima Publishing, 1997)

Pritikin, Nathan, *The Pritikin Program for Diet and Exercise* (NY: Grosset and Dunlap, 1979)

Pritikin, Robert, *The New Pritikin Program: The Premier Health and Fitness Program for the '90s* (NY: Simon and Schuster, 1990)

Robbins, John, *Diet for a New America* (Walpole, NH: Stillpoint Publishing, 1987)

Ruden, Ronald A., *The Craving Brain* (NY: HarperCollins, 1997)

Sauvage, Lester R., *The Open Heart: Stories of Hope, Healing, and Happiness* (Deerfield Beach, FL: Health Communications, 1996)

Slack, Warner V., *Cybermedicine: How Computing Empowers Doctors and Patients for Better Health Care* (San Francisco: Jossey-Bass, 1997)

St. James, Elaine, *Living the Simple Life* (NY: Hyperion, 1996)

—, *Simplify Your Life* (NY: Hyperion, 1994)

Time Life Books, *Dr. Koop's Self-Care Advisor* (NY: Health Publishing Group, 1996)

—, *The Medical Advisor: The Complete Guide to Alternative and Conventional Treatments* (NY: Time Life Books, 1996)

Vayda, William, *Mood Foods: The Psycho-Nutrition Guide* (Berkeley, CA: Ulysses Press, 1995)

Weil, Andrew, *Spontaneous Healing* (NY: Alfred A. Knopf, 1995)

Wurtman, Judith J., *The Serotonin Solution* (NY: Ballantine Books, 1996)

Yanker, Gary and Kathy Burton, *Walking Medicine* (NY: McGraw-Hill, 1990)

Web Sites

Alternative Medicine Home Page

http://www.pitt.edu/≈cbw/altm.html

Resource for finding information relating to alternative medicine

BetterHealth USA

http://www.betterhealthusa.com
   Information about food allergy blood testing and food sensitivities

British Diabetic Association

10 Queen Anne Street
London W1M 0BD
Tel: 020 7323 1531
Fax: 020 7637 3644
http://www.diabetes.org.uk

Center for Disease Control and Prevention (CDC)

http://www.cdc.gov/diabetes
   Diabetes research information translated for public education

Children with Diabetes

http://www.childrenwithdiabetes.com
   Online service for children and families with Type 1 diabetes

Consumer Information Catalog

http://www.gsa.gov/staff/pa/cic/food.htm
   Food and nutrition information from the US Consumer Information
Catalog

Diabetes Association

http://www.diabetic.org.uk

Eli Lilly

http://www.lilly.com/diabetes
General diabetes information from the Lilly Patient Education Program

Foot and Ankle Web Index

http://www.footandankle.com
General information about healthy feet and resources for foot problems

Health World Online

http://www.healthy.com
General information on health and wellness, nutrition

HealthComm, Inc.

http://www.healthcomm.com
Information on functional medicine, diabetes and anti-ageing research

Health*Max*, Inc.

http://www.healthmax.com
Portal to health and wellness information, programmes, products and services

ImmunoLabs

http://www.immunolabs.com
Information for health care providers about food allergy testing

Juvenile Diabetes Foundation International (JDF)

http://www.jdfcure.com
Juvenile diabetes educational information

Online Health Library

http://www.healthfinder.gov
 Online health information resources site

Vegsource

http://www.vegsource.org/lyman
 A source for vegetarian and vegan information

World Health Organization

http://www.who.ch
 Information on a variety of health topics and statistics

Worldguide Health and Fitness Forum

http://www.worldguide.com/Fitness/hf.html
 Information on anatomy, exercise and strength training (be sure to capitalize the F in Fitness in the website address)

Index

Quick and Easy Cooking for Diabetes

*Simple, healthy recipes for people with diabetes
and their families*

AZMINA GOVINDJI

Just because you have diabetes does not mean that you have to give up the pleasures of good food!

Published in association with the British Diabetic Association, and using the Association's latest dietary guidelines, this cookbook shows you how to prepare delicious, healthy meals with the minimum of fuss, for all the family.

Ranging from starters like Stuffed Mushroons and Baked Tomato and Olive Salad, through main courses such as Mediterranean Fish or spicy Chicken Pulau, to mouthwatering deserts like Banana and Chocolate Pie, this book is packed full of great recipes. Combined with tips and notes on healthy eating, this book is a perfect guide to how to looking after yourself and your health through eating the right foods.

Azmina Govindji, a State Registered Dietitian, was Chief Dietitian to the British Diabetic Association for eight years. She is the author of *Diabetic Entertaining* and co-author of *Recipes for Health: Diabetes.*

Recipes for Health

Diabetes

Low fat, low sugar, carbohydrate counted recipes for the management of diabetes

AZMINA GOVINDJI and JILL MYERS

This imaginative cookbook contains practical information about living with diabetes, as well as a delicious range of over 200 carefully devised and selected recipes for a diabetic diet. Published in association with the British Diabetic Association, all the recipes follow the latest dietary guidelines from the BDA.

It includes everyday meals for families, romantic meals for two, recipes for entertaining friends with, and much more. Each recipe is coded for calories and carbohydrate content, and are ideal for anyone wanting a healthy low fat, low sugar, high fibre diet.

For eight years Azmina Govindji was Chief Dietitian and Jill Myers was previously Home Economist for the British Diabetic Association.

Thorsons

Directions for life

www.thorsons.com

The latest mind, body and spirit news

Exclusive author interviews

Read extracts from the latest books

Join in mind-expanding discussions

Win great prizes every week

Thorsons catalogue & ordering service

www.thorsons.com